Early Success in Statistics

John Maltby and Liza Day *Sheffield Hallam University*

An imprint of **Pearson Education**

Harlow, England · London · New York · Reading, Massachusetts · San Francisco
Toronto · Don Mills, Ontario · Sydney · Tokyo · Singapore · Hong Kong · Seoul
Taipei · Cape Town · Madrid · Mexico City · Amsterdam · Munich · Paris · Milan

Pearson Education Limited
Edinburgh Gate
Harlow
Essex CM20 2JE

and Associated Companies throughout the world

Visit us on the World Wide Web at:
www.pearsoneduc.com

First published 2002   1004 853533 T

For more information about SPSS contact: SPSS UK Ltd, 1st Floor, St Andrew's House, West St., Woking, Surrey GU21 1EB
Tel: 01483 719200   Fax: 01483 719291

ISBN 0 130 19646 0

British Library Cataloguing-in-Publication Data
A catalogue record for this book is available from the British Library

10 9 8 7 6 5 4 3 2 1
06 05 04 03 02

Typeset in 9.5/12.5pt Stone Serif by 35
Printed in Great Britain by Henry Ling Ltd, at the Dorset Press, Dorchester, Dorset

For

our mothers

In memory of Bill Maltby,
who always emphasised the importance
of telling a story
when teaching mathematics.

Contents

Preface

Perhaps what the world needs least is another statistics/SPSS for Windows book. However, given the plethora of such books, it demonstrates how popular and really enjoyable statistics and statistical software are. Perhaps this last sentence is a little exaggerated. However, statistics can be a very exciting and inspirational topic. But it is a long journey and early success is very important if you really want to understand statistics. This book is designed to help you experience early success, by giving you working knowledge of a statistical package (SPSS for Windows), but it also aims to make demands on you to ensure that you have transferable skills for developing an understanding of multivariate statistics, and debates within the literature.

The title of this book is quite deliberate. It is designed for people who need to develop their knowledge and skills around statistics, but more importantly, it aims to build confidence through positive experience. It is important to stress that you have not just bought another statistics book, but you have bought a learning project. This project does not rely solely on the text you have bought. This learning project incorporates the SPSS for Windows program, the World Wide Web and other Windows applications.

Recent research into the ways in which students of statistics interpret success and failure in statistics (Maltby *et al.*, 1998; Onwuegbuzie and Daley, 1999), and in the ways in which they prefer to learn (Maltby, 2001) suggests the importance of quick success for many students. Statistics is taught and applied in a variety of ways. Hence you may be taught statistics in one way, and so may feel daunted or confused when you come across examples in class or research when a different method has been used. This book aims to be explicit about the different methods used. It will then compare different methods and provide you with the basic ideas underlying statistics and SPSS for Windows using a simple narrative.

When you have grasped the central concepts, you will move on to consider a strategic guide to understanding the different ways in which statistics are taught. This will develop confidence, enabling you to address issues underpinning more advanced thinking and research problems in statistics.

This progression will create a knowledge and skills base, giving you the confidence to persevere with statistics.

To build this knowledge and skills base we shall use techniques frequently employed by mathematics teachers to promote mathematical thinking, rather than teaching through explanations, rationale and formulae. There is recognition in academia of the need for skills books developing critical thinking for university students (e.g. Peck and Coyle, 1998, 1999). Such books try to exercise students' minds. With mathematical concepts, this can be done quite subtly, using mathematical puzzles. Similarly we want to develop your ability to adopt a 'statistical frame of mind'.

In summary, the book aims to provide a considered but friendly narrative introducing statistics and the numerous issues that surround each topic. It provides a unique combination of narrative, statistical knowledge, SPSS for Windows examples, logic puzzles, research papers, self-assessment tasks and a dataset to examine statistical concepts. Further, many of these statistical concepts will be presented to you using other media. You will be encouraged to develop your statistical knowledge and skills base through use of the Internet, with a website to accompany the book. Further, you also have at your disposal a range of PowerPoint animations to provide further ways of understanding statistics.

The learning package will build your statistical knowledge and skills base by:

- using text to introduce you to statistical concepts and SPSS for Windows in the context of research;
- allowing you to develop your knowledge and skills and to check your learning on the World Wide Web.

The text

The main body of the book is designed around a modular structure (one main topic per chapter). Rather than provide a full account of all possible statistical concepts and methods, the book aims to provide a comprehensive examination of the different ways in which statistical concepts can be explored. The book will deliberately *not* include the advanced statistics that are covered in other books, as these are typically taught on advanced research methods units, rather than on introductory units. The purpose of this book is to give a solid foundation and underpinning to fundamental statistical concepts.

The text is written purposely so that it is easy to understand. Concepts are simply explained and there is guidance on the application of this material within SPSS for Windows. A number of scenarios are used for tests that facilitate learning and develop transferable skills. Each chapter will contain exercises within the text to allow you to check your own learning and progress gradually towards understanding statistics in a research context. One important aspect of this is the Mathematical Energiser puzzles. These appear at the beginning of each section and are designed to energise your mathematical thinking for each session. Before attempting the chapter you have most probably done little mathematical work during the day. The aim of the energiser is meant to place you in the right frame of mind for the chapter. Further, these puzzles encourage you to practise using numbers in a fun context. Sometimes the puzzles will be designed to reflect some of the concepts or skills that are developed within a chapter, but these puzzles are also designed to develop your mathematical thinking skills.

Chapter 1, 'A map of statistical concepts', provides an overview of the different concepts to be presented in the ensuing chapters. It will include a map showing how different statistical debates and concepts are intertwined. This chapter will then show how specific components can be segregated to be examined in isolation. This segregation forms the basis of the following chapters. The individual components will then be combined to show, through a decision-making chart, how an overview of statistics can sometimes be reduced to a number of decisions.

Chapter 2 introduces you to one of the cornerstones of statistics, *variables*, and you will begin to use SPSS for Windows to develop good practice with regard to data input. Another central theme in this chapter is that you will be introduced to a dataset for use throughout the book. The dataset is fairly small, containing around 20 variables, and will be used to illustrate statistics and SPSS for Windows. However, the size of the dataset is designed to allow you to become familiar with the variables used very quickly.

Chapter 3 will focus on developing your knowledge and skills with *descriptive statistics*. It will introduce you to ways of 'describing' data. A central theme of this chapter is how to best use descriptive statistics and charts. You will not produce descriptive statistics and charts in isolation; you will be encouraged to understand that descriptive statistics and charts are used to enhance data, by identifying the best ones to use in each situation. So you will learn how to carry out these types of statistics by hand and by SPSS for Windows, and you will also learn how to apply this information in the context of an answer to a research problem.

Chapter 4 will introduce you to *inferential statistics*. This chapter introduces you to the concepts which underlie statistical hypothesis testing, with explicit reference to confidence intervals, distribution, probability, effects sizes and error. The chapter will explain the debate surrounding the use of parametric and non-parametric tests, by using a decision-making chart. Further, the decision-making chart introduced in Chapter 1 will be used to enable you to establish that there are two main families of tests: parametric and non-parametric.

Chapters 5 and 6 will introduce you to a number of statistical tests (Chapter 5: *parametric tests*; Chapter 6: *non-parametric tests*). For each of these tests, you will be taught the rationale behind it, learn under what circumstances the test should be used, and how to perform and interpret each of the tests by hand and via SPSS for Windows. However, the book will also help you to assimilate these concepts by seeing how they may be put into practice by research.

Chapter 7 will be used to consolidate your learning throughout the book. This chapter will include instructions for carrying out some independent statistical analysis on some existing datasets.

The chapters vary in length. In some cases you may get through a module in one session, while others may require two or more sessions. It is important that you work through each chapter and understand it fully before you proceed to the next one. There are some learning checks to allow you to see whether you should move on. Don't be afraid to repeat a module, or go back to one. The more familiar you get with statistics and SPSS for Windows, the better your learning will be.

The website

Please do use this. This website is well worth using again and again to help you again and again (www.booksites.net/maltby). These things are on the Web because we cannot put some things into the text. They will prove invaluable in building your statistical knowledge and skills base. You will find the following on the website.

Databases to use in Learning Check Exercises

You will be able to access databases to answer a number of research questions that will be posed in the book.

Multiple choice questions

You will be able to access multiple choice questions. There will be several banks of multiple choice questions to allow them to be used in three ways:

- to revisit ideas before moving on to new content;
- as learning checks at the end of each chapter;
- as a test of all concepts covered in the book to provide you with a check on your overall knowledge and learning.

Animations

You will be able to download a number of animations to allow you to explore the material in a different way. The aim of the animations is to provide you with a chance to see some of the operations or concepts 'in action'. These animations will support general statistical concepts and procedures in SPSS for Windows.

Message board

The website is meant to be truly interactive. There are pages you can use to input into the learning process. First, with a book like this, questions are bound to arise. We will try to provide answers to the most frequently asked questions to aid your learning. Second, we aim to develop a forum in which you can interact with other users, hold statistical discussions and exchange information. It is hoped that this in itself will become a huge resource, allowing sharing of knowledge about the use of statistics, statistical software packages and research-based issues.

Links

There are already many impressive 'statistics' websites. This page will provide links to the best of these sites.

Summary

The book's claim to bring success will be achieved by presenting a package that is simple to use, and book chapters that can be completed in sessions of 1 to 2 hours. As you complete a session you will be able to demonstrate that you have understood the concepts presented, by self-assessment and/or by accessing further problems/assessment via the World Wide Web. We hope you appreciate and use the different methods we have tried to employ to support your learning, with the aim to bring you early success. Please use all aspects of the learning, because we are confident that very quickly and with expertise, you will be able to develop a knowledge and skills base to allow you to use statistics with confidence.

References

Maltby, J. (2001) 'Learning statistics by computer software is cheating': Age as a factor for preference of the learning method of statistics. *Journal of Computer Assisted Learning* (also published on the internet: http://www.lancaster.ac.uk/users/ktru/jcalrn98.htm).

Maltby, J., Kirwan, J. and McCollam, P. (1998). Scores on a statistics test and scores of the Defence Style Questionnaire. *Psychological Reports*, **83**, 364–366.

Onwuegbuzie, A.J. and Daley, C.E. (1999). Perfectionism and statistics anxiety. *Personality and Individual Differences*, **26**, 1089–1102.

Peck, J. and Coyle, M. (1998). *Practical criticism*. London: Macmillan Press.

Peck, J. and Coyle, M. (1999). *The student's guide to writing*. London: Macmillan Press.

A Companion Website accompanies
Early Success in Statistics
by John Maltby and Liza Day

Visit the *Early Sucesses in Statistics* Companion
Website at www.booksites.net/maltby to find valuable
teaching and learning material including:

For Students:
- Study material designed to help you improve your results
- Multiple choice questions to test your learning
- Dataset for downloads in SPSS
- Discussion forum

For Lecturers:
- A secure, password-protected site with teaching material
- Downloadable Powerpoint slides
- A syllabus manager that will build and host your very own course web page.

Acknowledgements

We would like to thank Jill Wheeldon for reading and commenting on drafts of this book. Our thanks also go to our first year statistics classes for their continued input on how they approach and think about statistics. Thanks are also due to the anonymous reviewers.

We would also like to thank Dr Christopher Alan Lewis and Dr Christopher McConville for valuable advice over a number of years.

Permissions acknowledgements

We would like to thank those individuals and companies that have provided permission to reproduce material in this book, previously published elsewhere. A particular thank you to Dr Stephen Joseph for allowing us to reprint all the items from the Depression–Happiness Scale. If any permission has been overlooked the publisher will be pleased to make the necessary arrangement at the first opportunity.

CHAPTER 1

A map of statistical concepts

'Obvious' is the most dangerous word in mathematics.

(Eric Temple Bell, 1883–1960)

In this chapter:

- we provide an overview of the different concepts to be presented in different chapters, in order to provide a context for your learning;
- we show that statistics is a decision-making process;
- you begin to use some of the rules surrounding statistics.

Exercise 1: Chapter energiser

The aim of this chapter energiser is to begin to get you using numbers. The Domino Deal Puzzle (Figure 1.1) is straightforward and should take about 5 minutes. We have chosen this puzzle to help us illustrate how we are going to show an overview of different statistical concepts.

In this chapter we're going to tell you exactly what is going to happen in this book. Why? Because one of the problems with learning statistics is that beginners often don't have an overall picture of what they need to learn. This may lead to confusion: often concepts are learnt in isolation, without reference to the big picture. Liken this situation to knowing what happened at the end of a film. When you watch the film for a second time, you are able to appreciate plot developments, the significance of particular pieces of dialogue, and you notice events that seemed unimportant the first time, but on a second viewing you realise their importance. As you are unlikely to read this book again and again, we need to give the overview now.

Figure 1.1 Domino Deal Puzzle.

A set of dominoes has been laid out (using numbers instead of dots for clarity), but the lines that separate the dominoes are missing. Can you show where each domino in the set has been placed?

3	2	6	1	3	5	5
3	0	0	6	2	3	6
6	4	1	3	2	5	6
5	3	1	2	4	0	1
4	4	2	2	0	4	1
6	4	3	6	2	6	0
2	1	5	0	1	0	3
5	4	4	5	5	1	0

Check grid

A check grid has been given to help you, so each domino can be ticked off when it has been located.

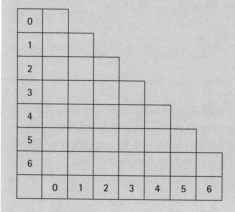

There are three stages to this:

- placing statistics within the proper research context;
- slotting in the different pieces of the statistics jigsaw to gain the overall picture;
- understanding statistics as a decision-making process.

Placing statistics within the proper research context

Statistics is part of a research process. At the beginning of any research project, there is always an idea that the researcher formulates in the shape of a question. For example, see Figure 1.2. The researcher will then collect some appropriate data to address the question. Statistics is one tool that the researcher may use to provide answers to the research question, i.e. statistics is one way to analyse data collected by a researcher. Therefore, it is important to view statistics as not an isolated skill, or an end in itself, but rather a way of answering a research question.

It is important to view statistics as a tool for understanding your data and answering your research questions. It is there to aid you in your research. You should not let your feelings about maths or numbers lead you to feel that statistics is in charge or central to the research process.

Statistics is a method of working out an answer to your research question using the data. We will use different elements of statistics as tools for carrying out statistical procedures, i.e. by hand-calculations and using SPSS for Windows software. Both these procedures work in similar ways and so you will gain an insight into how SPSS for Windows works.

Imagine at school, you were set a maths problem to which you had to find an answer. You would have a set of numbers, and you would have to carry out some procedure or apply some formula to those numbers to find the answer. Well, that is one way you might carry out statistics (this is known as 'hand-calculation').

However, you know that there are other tools to help you with your working out. The most popular one is the calculator, which helps you not to make mistakes, and quickens the process. SPSS for Windows is quite simply a very advanced calculator, and extremely useful. This is because it saves you time when you have lots of working out to do, can perform complicated formulae, and doesn't make mistakes. However, it works in exactly the same way as when you worked things out at school. It has a section where you use original numbers to set out the problem. The program has all the possible procedures/formulae that you could be asked to apply (in a very user-friendly way), and it has a section that provides you with an answer. We will introduce you to these different parts as we work through the book, but Figure 1.3 shows you which parts of SPSS for Windows correspond to each part of the process.

Figure 1.2   A research process.

Research idea – Collecting data – Analysing data – Answering research question

Figure 1.3 Comparison of the stages of mathematical working out, by hand, calculator and SPSS for Windows.

	Stage 1	Stage 2	Stage 3
By hand	The numbers given on a piece of paper	Apply the formula (perhaps add the numbers together)	Write out the answer
Calculator	Type in the numbers	Press + or − and =	Answer put on the screen
SPSS for Windows	This is called a data spreadsheet	SPSS for Windows program	Answer put in the form of OUTPUT

Slotting in the different pieces of a statistics jigsaw to gain the overall picture

There are a number of different parts to statistics. Each of these parts is important in providing an overall view. Therefore, we think a knowledge of statistics is a little like a jigsaw puzzle (see Figure 1.4a). By putting all the different parts together you can effectively apply statistics (Figure 1.4b). In this book we have used each chapter to address each piece of this jigsaw. The chapters represent all the outside blocks, starting with variables (Chapter 2) and moving clockwise to further issues (Chapter 7).

To place statistics in context, it is worthwhile summarising the main points of each chapter here. Don't worry if this seems overwhelming at this stage: we will address all these concepts in detail within each of the chapters. However, by surveying some of them now, and by getting you to use some of these ideas in the exercise in the next section, we can give you a real sense of the overall picture of statistics.

Figure 1.4a The separate parts of the Statistical Jigsaw.

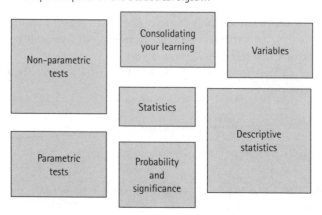

Figure 1.4b   How the separate parts of the Statistical Jigsaw come together.

Non-parametric tests	Consolidating your learning	Variables
	Statistics	Descriptive statistics
Parametric tests	Probability and significance	

Chapter 2: Variables

We will explain that variables are simply things that we measure that vary. Three issues arise from this:

- Researchers describe variables in two main ways: (a) categorical (variables are made up of separate categories, e.g. sex of a person); or (b) continuous (variables are numerical, e.g. amount of money a person earns).
- Each variable is made up of levels (for instance, sex is made up of two levels, male or female; variety of eye colour is made up of a larger number of levels, e.g. blue, brown, grey, green; while amount of money earned demonstrates numerical properties, with levels of variable increasing and decreasing, and the ability to add, subtract, multiply and divide using other numbers).
- Continuous variables can sometimes be described as demonstrating certain statistical properties that allow them to be used in *parametric* statistical tests. However, sometimes some continuous variables *do not* show these particular statistical properties, and when this happens, the variables are thought suitable to be used in *non-parametric* statistical tests.

Chapter 3: Changing, combining and describing variables

We will show you ways of examining and manipulating variables, using those techniques most often used by researchers. Also in this chapter we will introduce you to Descriptive Statistics, and the number of ways in which statistics can be best described by the researcher.

Chapter 4: An introduction to inferential statistics

We set the scene for using the tests described in Chapters 5 and 6, but involving concepts covered in earlier chapters.

Chapters 5 (Parametric tests) and 6 (Non-parametric tests)

We will outline a number of statistical tests (procedures/formulae) that are used within SPSS for Windows. These can be used to answer research questions you may have about the relationship between two variables. These chapters are the central core of your learning from this book.

Chapter 7

This chapter will be used to consolidate your learning throughout the book. It will include instructions for carrying out some independent statistical analysis on some existing datasets.

Statistics as a decision-making process

Ideas from the last section can be used to show that statistics can be viewed simply as a decision-making process. In the last section we made two distinctions regarding variables:

- that variables are viewed as either (i) categorical or (ii) continuous;
- that distinctions can be made between continuous variables in determining whether they show the properties which allow them to be used in (i) parametric tests or (ii) non-parametric tests.

In addition, we also pointed out that variables had different levels (sex is made up of two levels, male and female; eye colour is made up of a larger number of levels). Of some importance to statistics is a third distinction that is made when we use categorical variables: that a distinction can be made between categorical variables with (i) two levels and (ii) three or more variables.

These three distinctions are at the heart of statistics. In essence, the statistical test you use will depend on the distinction you use with your variables. Therefore, statistics is a decision-making process: you can answer questions about your variables in order to select the right statistical test to use in answering your research question.

We use Figure 1.5 'Decision-making table for choosing statistical tests' to show you how this overall decision-making process works. In Figure 1.5 you will notice a number of darkly shaded blocks with the words such as chi-square, Pearson, Spearman, related t-test, Wilcoxon sign-ranks test (to name some of them) written in them. These are the nine tests that we will cover in the book and are the end-points of the decision-making process. The aim of the decision-making process is to reach one of these end-points. To reach each point, what you must do is answer each of the questions

Figure 1.5 Decision-making table for choosing statistical tests.

Question 1 What combination of variables have you?	Which test to use	Question 2 Should your continuous data be used with parametric tests or non-parametric tests?	Which test to use	Question 3 How many levels has your categorical data?	Which test to use
Two categorical	Chi-square				
Two separate continuous	Go to Question 2	Parametric	Pearson		
		Non-parametric	Spearman		
Two continuous which is the same measure administered twice	Go to Question 2	Parametric	Related *t*-test		
		Non-parametric	Wilcoxon sign-ranks		
Two continuous which is the same measure administered on three occasions or more	Go to Question 2	Parametric	ANOVA (within subjects)		
		Non-parametric	Friedman test		
One categorical and one continuous	Go to Question 2	Parametric	Go to Question 3	2	Independent-samples *t*-test
				3 or more	ANOVA (between subjects)
		Non-parametric	Go to Question 3	2	Mann–Whitney U
				3 or more	Kruskal–Wallis

in each of the columns in turn, choosing your response from the rows below. What you must then do is continue along each row until you reach an end-point.

To illustrate this process let us use an example. Imagine we have Researcher A. Researcher A has measured individual anxiety levels and corresponding exam scores on a maths test. Researcher A has measured these variables to see if anxiety levels are related to exam performance. The researcher has decided that these are both continuous variables and, with the additional aid of using descriptive statistics, decided that the continuous variables fulfil the criteria for using with parametric tests.

Researcher A has two continuous variables that are separate, so you start on the second row. Moving along the line you must answer Question 2. Researcher A has decided, by using descriptive statistics, that both the continuous variables demonstrate statistical properties that suggest a parametric test be used. Therefore, the answer to Question 2 is 'Yes'. So you reach an end-point and you will use a Pearson statistical test. As you can see, the decision making that leads to choosing this test is fairly straightforward. The test itself will be explained in a later chapter.

Question 2 in the decision-making table

If there are two continuous variables, and one of the continuous variables is considered suitable for use in parametric tests, but the other is not, you use the non-parametric test. Therefore, you should answer Question 2 as 'No'.

You will also note that you don't necessarily have to answer all the questions to reach an end-point. In the last example, you didn't have to answer Question 3 to reach the answer, so don't always expect to answer all the questions.

Now you know about the three main issues that underlie statistics:

- Statistics, in its correct context, is part of a process; this part is a tool for answering a question.
- Using such a context is important for visualising how statistics knowledge comes together. Rather, it is best to view statistics as being made up of a number of parts forming an overall picture.
- Following this, you use these parts in the form of a decision-making process, to choose a particular statistical test with which to answer your research question.

Exercise 2: Using the decision-making table

Using the decision-making table, decide what test should be used by the following researchers:

- Researcher B has a measure of sex (males and females) and exam score on a maths test to see if men and women differ in their ability in maths. The researcher has decided that sex is categorical and exam score on a maths test is a continuous variable. Using descriptive statistics, Researcher B has decided that the continuous variable shows statistical properties that suggest a parametric test should be used (remember to identify how many levels each of the variables has).
- Researcher C has administered the same measures of anxiety before and after an exam to see if anxiety levels change before and after an exam. The researcher has decided that these are both continuous variables, but has found on this occasion, that both variables *do not* show statistical properties that suggest a parametric test should be used.

Answers to exercises

Exercise 1: Chapter energiser

Figure 1.6   Solution to Domino Deal Puzzle.

3	2	6	1	3	5	5
3	0	0	6	2	3	6
6	4	1	3	2	5	6
5	3	1	2	4	0	1
4	4	2	2	0	4	1
6	4	3	6	2	6	0
2	1	5	0	1	0	3
5	4	4	5	5	1	0

Exercise 2: Using the decision-making table

- Researcher B should use the independent-samples *t*-test.
- Researcher C should use the Wilcoxon sign-ranks test.

CHAPTER 2

Variables:
their central role in understanding statistics

The control of large numbers is possible, and like unto that of small numbers, if we subdivide them.

(Sun Tze (5th–6th century), Sun Tze Ping Fa*)*

In this chapter you will explore:
- what variables are, and how they are evident in many aspects of investigation;
- how variables are viewed by researchers;
- the form data takes to make exploration possible;
- use SPSS for Windows to create a datafile;
- the dataset that will be used as examples in the rest of the book.

Exercise 1: Chapter energiser

Try the following Number-cross Puzzle (Figure 2.1). It works the same as a crossword puzzle but just uses numbers. One skill developed in this chapter deals with putting in data using SPSS for Windows. This Number-cross Puzzle is a way of illustrating this skill.

Variables

Variables are quite simply things that vary. We are surrounded by variables. Examples of variables are: different types of events or objects, our feelings and attitudes, and other people's feelings and attitudes. Instances of these could be the different times that people get up in the morning, the different types of breakfast they might have (if any), the way that people feel about work, the number of hours they spend watching television at night, the

Figure 2.1 Number-cross Puzzle.

Answer the clues just like a normal crossword, only using numbers instead of words.
You can use a calculator if you wish.

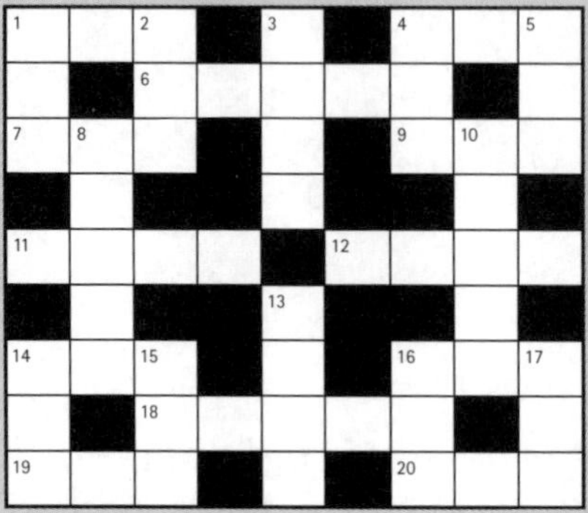

ACROSS

1. Subtract 11 from a dozen score.
4. Add 360 to 318.
6. Add the cube of 34 to the square of 52, subtract 6.
7. Treble 199, subtract 199.
9. Multiply the cube of 3 by 11.
11. Add the square of 87 to the cube of 6.
12. Double 4975, subtract 80.
14. Square of 29.
16. Average of 590, 626 and 716.
18. Add the cube of 27 to the square of 84.
19. Divide 8228 by 17.
20. Multiply the product of 4 and 9 by 7.

DOWN

1. Continue the sequence: 2184, 1092, 546, ...
2. Subtract 13 from the square of 31.
3. 12 342 divided by the square root of 36.
4. Subtract 219 from 841.
5. Add 158 to the square of 27.
8. Add 36 821 to 55 893.
10. Add 38 466 to 61 238.
13. Add 3579 to 3398.
14. Multiply 189 by 5, subtract a fifth of 355.
15. Divide 39 483 by 321, and add 1.
16. Divide 204 832 by 296.
17. Square 22, subtract 2.

time they go to bed. These examples are slightly flippant, albeit true. As researchers we tend to be interested in those variables related to our discipline. Therefore, a psychologist may be interested in levels of depression, scores on a personality test, IQ level; a nurse may be interested in recovery rates of patients, different types of illness, and different types of treatment; a sociologist might be interested in social class, level of alienation, and work roles; and an economist would be interested in interest rates, unemployment figures, and levels of supply and demand. Whatever the area of social science or health, variables are at the heart of research, and therefore, are at the heart of statistics.

Identifying variables

One important skill that researchers must have is to be able to identify accurately variables which exist in an area of research. We are going to spend a little time looking at how to develop the skill of identifying variables.

Read the article 'Scientists warn of 30% rise in human BSE' by James Meikle (2000) in *The Guardian* (Figure 2.2).

Figure 2.2 'Scientists warn of 30 per cent rise in human BSE'.

Scientists warn of 30 per cent rise in human BSE
What's wrong with our food? – special report

James Meikle
Tuesday, 18 July 2000

Government scientists yesterday warned of a sharply accelerating trend in the incidence of human BSE after studying the pattern of the disease so far.

They said the number of reported cases may in fact be rising at between 20 per cent and 30 per cent a year despite the apparently varied annual death rates over the past five years.

The prediction came as it was revealed that the death toll from the incurable condition officially known as vCJD had risen by a further two in the past fortnight to a total of 69, and 14 so far this year.

The scientists said that there was now a 'statistically significant rising trend' in the number of victims since the first casualties first displayed signs of the disease in 1994, although it was still too early to forecast the ultimate number of deaths caused by vCJD.

This year's toll is already equal to that for the whole of last year when the number dropped. A further seven people still alive are thought to be suffering from the condition. The scientists have come to their conclusion about the progress of the disease after analyses of monthly figures, including studying the dates at which friends, relatives or doctors first noted symptoms.

The period between this and eventual death has varied between 7 and 38 months, with an average of 14 months, although the incubation period before symptoms become evident is believed to be several years longer.

Stephen Churchill was the first known death from the disease in May 1995, although it was not formally identified or officially linked to the eating of beef in the late 1980s until March 1996. Three people died in 1995, 10 in 1996, 10 in 1997, 18 in 1998 and 14 last year.

Members of the government's spongiform encephalopathy advisory committee took the unusual step of publishing the figure immediately after their meeting in London yesterday because of the recent interest in a cluster of five cases around Queniborough in Leicestershire.

These included three victims dying within a few of months in 1998, a fourth who died in May and another patient, still alive, who is thought to be suffering from the same disease.

The scientists said this was 'unlikely to have occurred by chance but this cannot be completely ruled out' and they would be closely informed about local investigations. The Department of Health last night said it could not elaborate on the significance of the new analysis until ministers and officials had considered the scientists' new advice.

The figures came amid reports that sheep imported by the US from Europe were showing signs of a disease which could be linked to BSE in cattle. Government scientists are to hold talks with their US counterparts after the US agriculture department ordered the destruction of three flocks of sheep which were in quarantine in the state of Vermont.

(Reproduced with permission of author and publisher from Meikle (2000). © *The Guardian*, 2000)

You will see that there are many variables that government scientists are interested in. At one level it may seem that scientists are only interested in the levels of vCJD, and in how many people have died of vCJD. However, there are other variables that can be identified within this article:

- the year in which people have died (to consider trends in the disease);
- changes in the frequency of vCJD, by looking at changes from one year to the next;
- the time period between which 'friends, relatives or doctors first noted the symptoms' and eventual death, which varies from 7 to 38 months;
- whether people have died of vCJD, or another related disease;
- where the vCJD case occurred; here there is an emphasis on Queniborough in Leicestershire.

We can see, therefore, that even within a fairly straightforward area of research, many variables emerge during the course of an investigation.

A useful thinking skill that you can develop is to be able to identify what possible variables are contained within a research area. Read the next article by Kirsty Scott (2000) which also appeared in *The Guardian* (Figure 2.3). Try to identify, and list in the box below, as many variables as you can that you see emerging from this report. You should be able to name a number of variables.

Identifying variables within academic titles and text

What is particularly interesting about the last example is that the report begins to speculate about some of the causes of lung cancer. The researchers (in the report mentioned) suggest that a number of different variables have contributed to lung disease. These include whether workers have worked excessive hours or not, the number of times they may have worked excessive hours, and whether protective equipment is worn.

Also, being able to identify how variables relate to other variables is central to any research. All researchers are interested in asking, and trying to answer, research questions about the relationships between variables. Some researchers will refer to research questions in different ways. Some researchers refer to hypotheses, aims, or objectives, but basically these are terms used to answer a question about research. Therefore, a research question a health professional might ask is whether smoking (Variable 1: whether a person smokes, or not) is a cause of heart disease (Variable 2: whether a person develops heart disease, or not). A psychologist might ask whether being optimistic about life (Variable 1: your level of optimism can be determined by using a questionnaire) helps decrease your depression (Variable 2: your level of depression can be determined by using a questionnaire). So far, we have used examples where researchers may be seeking to establish connections between variables.

Figure 2.3 'Miners' long hours blamed as lung disease returns'.

Miners' long hours blamed as lung disease returns

Kirsty Scott
Tuesday, 18 July 2000

Miners at a Scottish colliery are suffering from a serious lung disease that health experts thought had been virtually eradicated.

Routine tests have found that nine miners at the Longannet colliery in Fife have developed pneumoconiosis, or black lung, which is caused by inhaling coal dust. A further 11 have abnormalities in the lungs, an early stage of the condition. The condition, which can lead to debilitating and sometimes fatal respiratory disease, was thought to have almost disappeared with the introduction of new safety and screening measures in the mid-1970s.

Last year a compensation scheme was agreed for miners affected by the disease after the biggest ever personal injury action in the UK. A Health and Safety Executive report on the Longannet findings is expected to blame excessive working hours and a failure to use protective equipment properly.

Dan Mitchell, HSE chief inspector of mines, said it was unusual to have found such an outbreak. 'But certainly in recent years the number of workers in mines attending for x-ray has been falling,' he said. 'It's not as good as it used to be and we only know about the prevalence of disease from the people who are x-rayed.'

The re-emergence of the disease has also surprised medical authorities at the Scottish pulmonary vascular unit at the Western Infirmary in Glasgow. 'I am really quite surprised because we have known about this condition for years and screening measures have been in place for years,' said the unit head, Andrew Peacock. 'We know what causes it. We expect old cases from the past but new cases coming along now does surprise me.'

Under regulations introduced in 1975 miners are only supposed to work 7-hour shifts, but many work overtime. They are also expected to have lung x-rays every five years, but at Longannet only around 70 per cent of men took part on the last occasion.

Representatives from the National Union of Mineworkers met HSE officials yesterday to discuss the situation. Peter Neilsen, vice-president of the NUM in Scotland, said: 'We thought that disease had disappeared. As a union we are concerned and it is our intention to take stock of the situation.'

The Scottish Coal Deep Mine Company, which runs Longannet, issued a statement saying that health of employees was of the utmost concern. More than 82 000 claims for compensation have been filed since the miners won their health case against the government and the nationalised coal industry. They claimed it had been known for decades that dust produced in the coal mining process could cause diseases like emphysema and chronic bronchitis and that not enough was done to protect them.

(Reproduced with permission of author and publisher from Scott (2000). © *The Guardian*, 2000)

Variables: Write your answers in here for the exercise above.

However, it is also worth noting that researchers are sometimes equally interested in not finding relationships between variables. For example, a health researcher would be interested in ensuring that a new drug does not have any major side effects.

Exercise 2: Identifying variables

Using the captions from newspaper articles given in Figure 2.4, try to identify the variables, and what possible links/or connections the journalist and researchers are trying to identify and establish.

Figure 2.4   Identifying variables from newspaper article headlines.

1.   Overfishing and global warming land cod on endangered list (S. Brown, 20 July 2000)
     _____
     _____

2.   Big spend may bring [inflation] rate rise, says MPC (C. Denny, 20 July 2000)
     _____
     _____

3.   Conveyancing website aims to cut gazumping and stress of moving home (A. Chrisafis, 20 July 2000)
     _____
     _____

Now try it with some academic journal titles (Figure 2.5).

Figure 2.5   Identifying variables from journal article headlines.

1.   Symptoms of anxiety and depression among mothers of pre-school children: effect of chronic strain related to children and child care-taking (Naerde *et al.*, 2000)
     _____
     _____

2.   Characteristics of severely mentally ill patients in and out of contact with community mental health services (Barr, 2000)
     _____
     _____

3.   Gender and treatment differences in knowledge, health beliefs, and metabolic control in Mexican Americans with type B diabetes (Brown *et al.*, 2000)
     _____
     _____

Distinctions *within* variables

So, now we know how to identify a variable, we now need to understand that certain distinctions are made within variables. These distinctions are called levels of a variable, and are quite simply the different elements that exist within a variable. Therefore, all variables have a number of levels. Sex has two levels: you are a man or a woman; age has numerous levels ranging from 1 years old to a probable maximum of 120. We have seen levels in all the examples we have used so far. Take, for example, the variables we identified in the vCJD article:

* *The year people have died.* The levels here are the years, 1992, 1993, 1994, etc.
* *The time period between which 'friends, relatives or doctors first noted the symptoms' and eventual death, which varies from between 7 and 38 months.* The levels here are in months.
* *Whether people have died of vCJD or another related disease.* The levels here could be two-fold. The first version might only contain two levels: whether the person had died of vCJD; or whether the person had not died of cVJD, but of a related disease. A second version might contain levels that describe each related disease leading to many more levels.
* *Where the vCJD case occurred.* Here there is an emphasis on Queniborough in Leicestershire. Levels here are different places, e.g. Leicestershire, Nottinghamshire, Derbyshire.

Distinctions *between* types of variables

You now know you can make distinctions *within* variables: we now need to expand on these distinctions and understand that researchers then go on to make distinctions *between* different types of variables. The main reason for these distinctions is that they underpin the choices that need to be made when using a statistical test. Mainly, there are two common sets of distinctions that researchers make between variables. These distinctions rely on how researchers view the levels that exist within a variable.

Set 1: Distinctions between nominal, ordinal, interval, and ratio

In this set, the first type of variable is called *nominal* and this entails merely placing levels into separate categories. The levels of this variable type (a nominal variable) are viewed as unique from one another. For example, the sex of a person has two levels, male and female. As sex is biologically determined, individuals fall into one category or another.

The next type of variable in this set is called *ordinal*, and this entails placing the levels into ranked ordered categories. We often see this type of variable in TV or movie magazines for ratings for films (4 = superb, 3 = good, 2 = OK, 1 = awful).

The next types of variable are *interval* and *ratio* and the levels for both these variables are numerical, meaning that they comprise numerical values. This means that the levels of these variables are numbers. They do not represent something else, such as the numbers used above for the ordinal variable, for rating a film. However, there is a distinction between these two types of variable. Ratio data has an absolute zero, i.e. it can have an absence of the variable, while interval data does not have an absolute zero. So, for example, the number of children in a family is a ratio variable because families can have no children.

Interval variables do not have an absolute zero. Common examples of interval variables are many of the psychological constructs we use, e.g. self-esteem. The measurement of self-esteem is not readily available to us (as opposed to simply counting the number of people in a family). Measurement of self-esteem in research will normally involve adding together the responses to a number of questions to produce a self-esteem score. As such, we only refer to self-esteem in terms of low or high self-esteem, or relative terms, such as a person having higher, or lower, self-esteem than another person. Because of this type of measurement we can, at no point, establish that there is an absence of self-esteem (i.e. has an absolute zero), so researchers treat many scales (particularly of psychological constructs) as interval data.

Set 2: Distinctions between categorical, discrete and continuous

The other set of distinctions are those that are made between *categorical, discrete* and *continuous* variables. Categorical data is the same as nominal data. Yet in this set of distinctions, researchers make an important distinction between discrete and continuous data. Here, both these variables are numerical. However, continuous variables allow for decimal points (1.23, 1.67, 10.567 865 67), while discrete data does not allow for decimal points (e.g. 1, 2, 3, 4, 5).

As you can see from these two sets of distinctions between types of variables, there are different ways that researchers make distinctions between variables (and there are also a number of ways that people merge these different sets of definitions and develop different understandings). Having different definitions can be terribly confusing for people starting out on statistics. This is sometimes particularly difficult, as teachers of statistics will adopt different definitions. If you are on a course you will probably, in time, come across different teachers who use different ways of defining variables.

To put it simply, one of the main sources of possible confusion is how teachers of statistics differ in the way they view the ordinal variable (set 1) and the discrete variable (set 2). There are two possible ways in which researchers perceive both types of variables. For many researchers, both ordinal and discrete variables are essentially ordered in a numerical way, and as such they believe that they should be viewed in the same way as continuous/interval/ratio variables.

However, for other researchers, ordinal and discrete variables represent separate, unique levels that do not represent numbers on a continuum. One example often cited, to support the latter point, is the distance represented *between* the levels of an ordinal variable. Unlike numerical variables, in which the distance between levels is equal (the distance between 1 and 2, and 2 and 3, and 3 and 4, are the same, '1'). For ordinal data, the distances between the levels are not the same (for example, in the example of the film ratings above, the distances between superb, good and OK may not be of equal value).

However, there is a simple strategy (see Figure 2.6) with which to tackle this issue without confusion, when beginning statistics (though this is by no means foolproof but provides a useful survival technique). This strategy relies on emphasising the similarities between the two descriptions, rather than the differences. Within this strategy, variables can instead be treated as either (i) categorical (distinct levels) or (ii) continuous (numerical ordered levels). This means that, in the first place, you should treat nominal/categorical variables as merely categorical, and ordinal/interval/ratio/discrete/continuous variables as merely continuous. However, you must always remember that different researchers treat ordinal and discrete variables differently. Some researchers insist these variables are categorical and some insist these variables are continuous. As such, there is no right or wrong view. Rather, you just have to be aware that this distinction occurs. This distinction has some implications for choosing which statistical test to use; however, we will return to this issue in Chapter 4 when an overall guide to the decision-making process in choosing which statistical test to use will be presented. However, the employment of this strategy will allow you to navigate through the statistics contained in this book.

Figure 2.6   Simple strategy for remembering distinctions *between* variables.

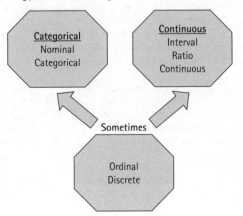

Exercise 3: Identifying more variables

Using the article 'US elections prove a TV turn-off' (Figure 2.7), identify the variables (and the possible levels within the variables). For the variables you identify, note whether the variable is categorical or continuous. If you are feeling confident you may even make further distinctions between types of variables (i.e. is the variable nominal/categorical, ordinal, interval, ratio, discrete, continuous).

Figure 2.7   'US elections prove a TV turn-off'.

US elections prove a TV turn-off

Martin Kettle
Friday, 21 July 2000

Fewer Americans than ever before are likely to watch this year's party conventions on television, which may mean fewer people at the ballot boxes too, writes Washington correspondent Martin Kettle.

For years, the central axiom of modern campaigning has been that elections are won and lost on television. So, it must say something about the state of American politics that fewer people than ever are expected to watch this summer's two party conventions on television than ever before.

It is also ironic that America's national television networks are busy scaling down their coverage of the upcoming conventions to the lowest in years.

Four years ago in San Diego, the Republicans ran a slick, controlled convention from which every possible negative image was ruthlessly expunged. Journalists began moaning that what had once been a fibrous and newsworthy part of the political process was being turned into an 'infomercial'. The ABC network anchor Ted Koppel even pulled out of San Diego after two days in protest, taking his crew with him.

When the Republicans gather in Philadelphia on 31 July for their 2000 convention, the television networks will be back. But only just.

Until this week, when they were forced to change their plans, ABC had been intending to cover the opening night of the convention only in the halftime interval of a pre-season American football game, with longer reports on the other three nights. CBS, which had been planning to cover just two of the four nights at Philadelphia, said on Monday that it now intends to offer at least some coverage every night.

But the bottom line in Philadelphia – and in the 14–17 August Democratic convention in Los Angeles too – is that things ain't what they used to be. In the 1970s, the networks were there every night offering extensive live coverage. In 1976, a peak year, the conventions had a combined rating of 35 points, meaning that 35 per cent of all American households were watching the conventions in an average broadcasting minute.

By 1996, the rating was down to 24 points, the lowest yet recorded. The average family watched less than four hours during the two conventions last time round. No one expects that record to survive the 2000 convention season.

continued

Figure 2.7   (*continued*)

Part of the reason for the likely slump this year is the revolution in television news. While the networks stay away, or scale down their coverage, their cable subsidiaries and their competitors are actually increasing their broadcast hours. MSNBC will in effect take over political coverage from its parent NBC. CNN will offer nearly 100 hours coverage from each of the conventions.

But the mathematics of such flexibility points steadily downwards. CNN may hope to put 10 per cent on its audiences in convention weeks, but the cutbacks on the networks – which reach many more households than the cable channels do – mean that the overall numbers of convention viewers will be the smallest ever.

This decline is matched by a downturn in interest. Harvard University's Kennedy School of Government reported this week that 43 per cent of voters in a poll for their 'Vanishing Voter Project' said they did not plan to watch any of the Republican convention, while 38 per cent didn't expect to tune in to the Democrats. Four years ago – when the record low was set – the corresponding figures were 23 per cent and 21 per cent.

One significant consequence is that both parties will feel driven to compensate by buying more advertising time. To pay for the increase in slots, the campaigns will have to raise even more money, including the 'soft money' that so many people feel is debauching the electoral process. The break-up of the old media monopolies is increasing the power of the rich donors, another paradox.

The parties have invested a lot of time and muscle in trying to get more coverage for their conventions. The changes of heart by both ABC and CBS about their coverage only came after the Democrats scheduled Bill Clinton for an opening night address in LA, forcing the networks to cover the president and compelling them – on grounds of balance – to also cover General Colin Powell's opening night speech to the Republicans.

But the parties also know that the process of change is irreversible, and they are adapting to it. Both the conventions will be available live on the internet, as the parties try to reach out to more voters. The Republicans are inviting net surfers to become 'dot.com delegates', while the Democrats are promising 'E-mersion coverage'.

From a media point of view, coverage of the 2000 conventions will be the most innovative yet. But the sober fact, in spite of all the new media, is that viewing will slump overall. It is not surprising that most forecasters think that, come election day on 7 November, this lack of contact will ultimately mean that fewer people will vote this year than in any presidential election on record.

(Reproduced with permission of author and publisher from Kettle (2000). © *The Guardian*, 2000)

Wider context of data-collecting research and SPSS for Windows

The next stage in this process is to begin to convert variables into usable data to answer a research question. This follows a process:

1.   Collect the data.

2.   Code the data.

3.   Enter into a data sheet on SPSS for Windows.

1. Collect the data

To collect variable data you would usually administer questions to a number of people. This could be done in a number of ways. The most common in statistics would be via a questionnaire, experiment, or one-to-one interview. The different ways in which you can collect data is a subject more for research methods than statistics books. What we are concerned with here, is how you begin the analysis of the data. However, in order to demonstrate this stage, we need to provide some data. We are going to use data that we have generated from four people, who answered the questionnaire given in Figure 2.8.

Figure 2.8   The Questionnaire X survey.

Questionnaire X

Please answer these questions by circling the appropriate response or filling in the gap.

Question 1: Are you   Male   or   Female?

Question 2: What colour eyes do you have?   Brown   Blue   Green   Other

Question 3: What age are you?_____

Thank you

Four people answered these questions for us. The first person (respondent 1) was male, had brown eyes and was 29 years old. The second person (respondent 2) was female, had blue eyes and was 32 years old. The third person (respondent 3) was male, had blue eyes and was 25 years old. The fourth person (respondent 4) was male, had green eyes and was 28 years old.

2. Coding the data

First, we must convert all the variables and their separate levels into numbers. This happens because all statistical packages tend to use number codes to deal with variables because they are mathematical machines. Researchers code all variables into numbers regardless of whether they are numerical or not. This should not intimidate you. You simply decide what the number codes are for each variable. So, for our example, where we have asked respondents what gender they are, they indicate whether they are male or female. You would then, in preparing the data for coding, assign a code for each category. Usually this would be '1' for Male, '2' for Female (or vice versa). We would repeat this process for all the other variables. So, for the 'colour of eyes' variable, we would code Brown as 1 (note you can use the same numbers for different variables), Blue as 2, Green as 3, and Other as 4. It is worth noting that it is not worth coding the final question about age into numbers, as each response is already a number.[1]

3. Enter into a data sheet on SPSS for Windows

An important skill in this part of the analysis process is to be able to visualise the way that data is laid out. To make this skill explicit, we are going to use an example on paper (the next section in this chapter will show how to do this on SPSS).

Data is laid out in SPSS for Windows in a grid format (Figure 2.9). In this grid the columns represent different variables, and the rows represent each person (respondent). What you aim to do then is to fill the grid with all the information you have gathered.

Figure 2.9   A data grid for the Questionnaire X survey.

	Sex	Colour of eyes	Age
Respondent 1			
Respondent 2			
Respondent 3			
Respondent 4			

Let us use the data collected from the four people on their sex, eye colour, and age. Respondent 1 was male, had brown eyes and was aged 29; Respondent 2 was female, had blue eyes and was 32; Respondent 3 was male, had blue eyes and was 25; and Respondent 4 was male, had green eyes and was aged 28.

Using the codes that we assigned in '2. Coding the data' (sex (male = 1, female = 2), colour of eyes (brown = 1, blue = 2, green = 3, other = 4) and age (leave as actual number)) we would assign the following code for each person:

- Respondent 1: sex = 1, colour of eyes = 1, age = 29
- Respondent 2: sex = 2, colour of eyes = 2, age = 32
- Respondent 3: sex = 1, colour of eyes = 2, age = 25
- Respondent 4: sex = 1, colour of eyes = 3, age = 28

---

[1] Please note that for some reason people try to always put age into categories, 18–25, 26–30, 31–50, to probably avoid asking people directly because they feel it may be insensitive. These sentiments are laudable, but try to avoid splitting age in this way. Age is clearly a continuous variable. Also, if you need to categorise it later you can condense this variable into categories in later data analysis.

Figure 2.10 Completed data grid for the Questionnaire X survey.

	Sex	Colour of eyes	Age
Respondent 1	1	1	29
Respondent 2	2	2	32
Respondent 3	1	2	25
Respondent 4	1	3	28

These would then be transferred on to the grid, as in Figure 2.10, putting each number in the appropriate box, according to where it lies within the columns and rows. So to illustrate, the highlighted box with '1' in it in Figure 2.10 shows that Respondent 1 was male.

This is really the essence of how data is inputted and presented in SPSS for Windows. In the next section we will show you how to do this in SPSS for Windows. However, before we do this, try the following exercise.

Exercise 4: Using a dataset

Using the coding for sex, colour of eyes and age above, answer the following questions for the dataset in Figure 2.11.

Figure 2.11 Data grid for Exercise 4: Using a dataset.

	Sex	Colour of eyes	Age
Respondent 1	2	1	32
Respondent 2	2	4	35
Respondent 3	2	2	21
Respondent 4	1	2	34

1. What colour eyes does Respondent 2 have?
2. How old is Respondent 2?
3. How many females are there in the present sample?
4. How many people are over 30 in the present sample?

Inputting data into SPSS for Windows

When you start up SPSS for the first time, the first 'window' that you see will be SPSS's data grid (Figure 2.12), which, in effect, is a spreadsheet. You use this sheet to input your data into SPSS, and it is very similar to the layout of the grids we used in the last section. In this section we will learn how to type data into SPSS.

Before typing in data we need to do two things: (1) collect the data to type in; and (2) prepare the SPSS file for data entry.

1. Collecting data

For this example we collected data from five respondents who answered the questionnaire in Figure 2.13. The five respondents answered the three questions as shown in Figure 2.14. We now have our data, so let us input this data into SPSS for Windows.

Figure 2.12 The SPSS Data Editor.

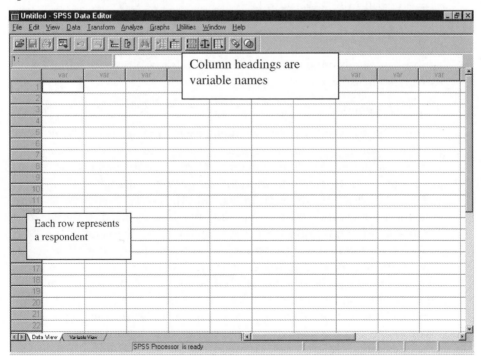

Figure 2.13   The Anxiety Questionnaire.

Anxiety Questionnaire

Please answer these questions by circling the appropriate response or filling in the gap.

Question 1. Sex: Male   Female

Question 2. What age are you?_____

Question 3. How anxious would you describe yourself?
          (1) Not at all   (2) A little   (3) Not certain   (4) Quite a lot   (5) A lot

          Thank you

Figure 2.14   Breakdown of response by the five respondents for the Anxiety Questionnaire survey.

Respondent 1	Respondent 2	Respondent 3	Respondent 4	Respondent 5
Q1. Male	Q1. Female	Q1. Female	Q1. Male	Q1. Male
Q2. 25	Q2. Aged 27	Q2. 33	Q2. 29	Q2. 21
Q3. A little anxious	Q3. Not at all	Q3. Anxious a lot	Q3. Not certain	Q3. Anxious a lot

2. Preparing the SPSS datafile

Before typing in the data, it is considered good practice to label the variables in the dataset. This comprises three stages: giving the variables names; if appropriate, identifying what each of the levels mean for each variable; and identifying missing values.

Variable names

Within SPSS for Windows we can name each variable. The variable name by convention is short (up to eight characters) and is used to remind you which variable each column represents. Giving variable names to our variables above is fairly straightforward for our first two variables, sex and age of respondents. However, for our third variable, the level of anxiety of respondents, this is longer than eight letters and, therefore, we need to abbreviate it. On this occasion it is fairly simple. We could just call the variable 'Anxiety'. However, for future use, it is useful to know that SPSS will not allow certain characters (such as @, #), or spaces between words, to be used when creating a variable name. Notwithstanding these formatting issues, let us name our three variable names in the SPSS file. On the front data sheet, click on the **Variable View** button in the bottom left hand corner (Figure 2.15).

You are now in the sheet that allows you to input information about each of your variables (Figure 2.16). Though there are many columns here we are

Figure 2.15   The Variable View button.

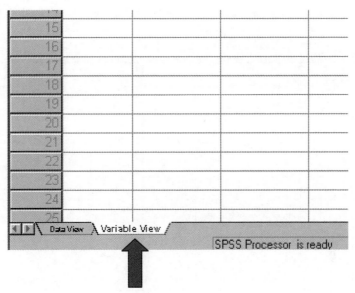

Figure 2.16   The Variable View data grid.

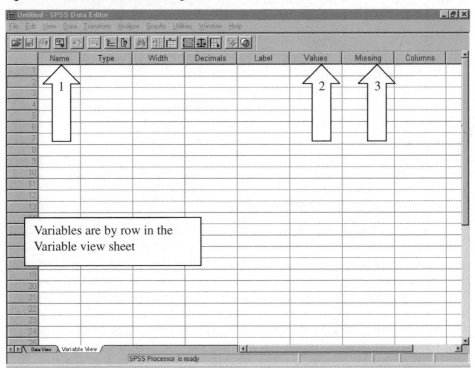

Figure 2.17   Variable View window: SPSS Data Editor.

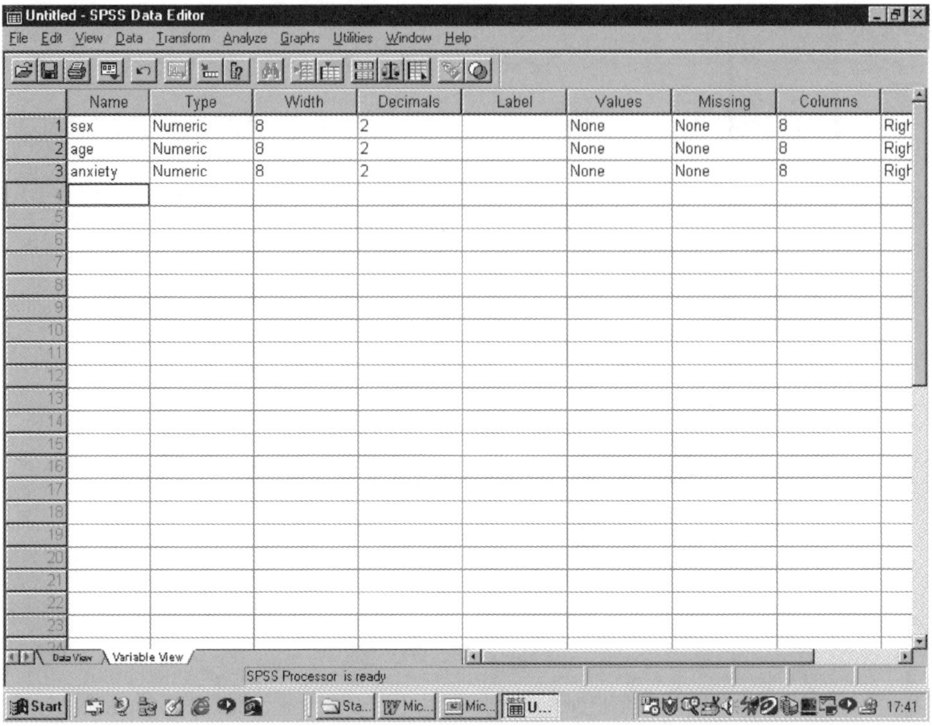

only interested in using three of these columns: (1) Name; (2) Values; and (3) Missing.

Now to type our variable names in, all you do is type your variable names in the 'Name' column, with each row representing a different variable (Figure 2.17). Here, type in sex, age and anxiety into the first three rows of the first column. After you have typed in each word, press the arrow keys to move into the box you want to. Here, it will be the down arrow key each time (alternatively you can position the cursor using the mouse).

Labelling each level of the variables

For each variable, numerical values are assigned to each level. For some of your data you don't need to be reminded what each level means. For example, for age you will be able to remember that each number represents years. However, for other variables where you have assigned numbers (for example, where we have done it for sex of a respondent, 1 = male; 2 = female) it is useful to record these on the data sheet. To do this in SPSS for Windows, click on the box under 'Values' that corresponds to the variable you wish to label, and then click on the grey section that appears (Figure 2.18). You should now have a box that looks like Figure 2.19.

Figure 2.18   Close up of the variables inputted in the Variable View window.

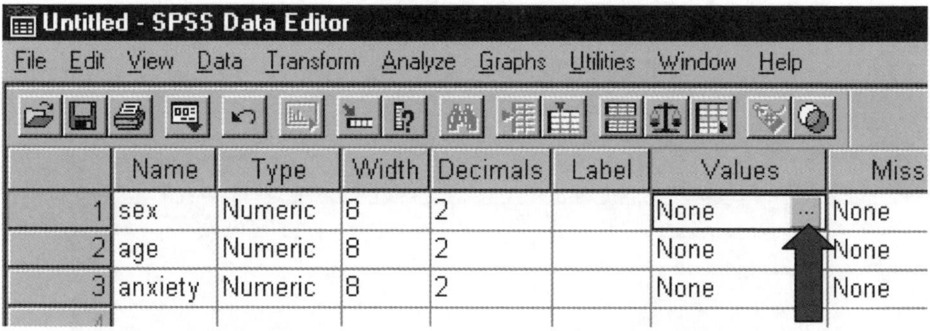

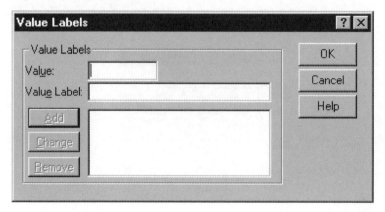

Figure 2.19   Value Labels window.

Here for Sex:

1. Type 1 into the **Value:** box (for the first level).
2. Type Male into the **Value Label:** box.
3. Press the **Add** button.
4. Type 2 into the **Value:** box (for the second level).
5. Type Female into the **Value Label:** box.
6. Press the **Add** button.

You should now have a box that looks like Figure 2.20. Then press the OK button. Your values column should look like Figure 2.21.

Value labelling other variables

We would miss out the Age variable, as we know what each level means (each number represents age in years). However, for the Anxiety variable, we have assigned a number for each response: (1) not at all; (2) a little; (3) not

Figure 2.20   Value Labels window with our value labels inputted.

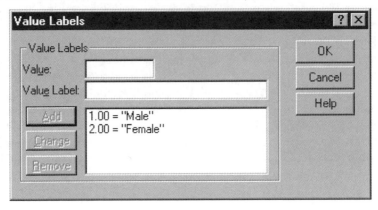

Figure 2.21   A completed **Values** box.

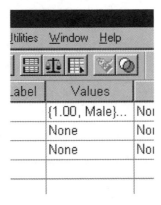

certain; (4) quite a lot; and (5) a lot. Using this data sheet, input the value labels in the values column for the anxiety variable.

Missing data

There is also a procedure to allow you to deal with data when people have not given you the required information. There are many reasons for why someone might not have answered a question: they may not have wanted to as they felt it too personal, or if it was on a question-naire, they may have missed it by error. Either way SPSS allows you to deal with this missing data by assigning a value to the data to tell the computer to treat that data as missing, rather than treat it as a number. You can choose what number you assign to it, but traditionally the number '9' is used with single figure values, or if 9 is already used then 99, or 999.

There is no need to input this data with the current data because we have no missing values. However, if you want to practise this it is fairly straightforward. In SPSS for Windows, it is similar to labelling. Let us assign a missing value for the Sex variable. Click on the box under **Missing** that corresponds with the Sex variable and then click on the grey section that appears. You will then get a box that looks like Figure 2.22. Click on the button next to **Discrete missing values** and type in your values. Here we've put in 9, but as you can see, SPSS for Windows allows you to put in a number of values, or in the row below a range of values (these options can be used for advanced options). You then press OK. You should now see a 9 under the missing column in the Sex variable row. Now, when you have to allocate a missing value to a missing response to the Sex item you just type in 9, and the computer will ignore it.

Figure 2.22   Missing Values window.

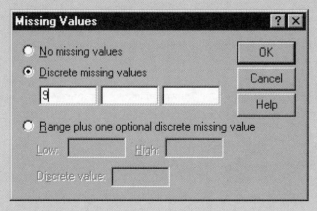

Exercise

1. Repeat this procedure for the Anxiety variable.
2. Imagine some respondents did not want to reveal their age to you. What number would you assign as a missing value?

3. Putting your data into SPSS

Now we are going to put the data in. To move from the **Variable View** data window to the **Data View** window, press the **Data View** button in the bottom left hand corner. Notice how the first three columns all have variable names. Now you can start putting your data in. Remember the Anxiety Questionnaire (Figure 2.23) and responses (Figure 2.24).

Remember how you presented the data in the paper example. It is exactly the same as that. For the first row (respondent 1) you would put '1' in the first column, '25' in the second column and '2' (a little) in the third column (Figure 2.25).

Figure 2.23 The Anxiety Questionnaire.

Anxiety Questionnaire

Please answer these questions by circling the appropriate response or filling in the gap.

Question 1. Sex: Male   Female

Question 2. What age are you?_____

Question 3. How anxious would you describe yourself?
(1) Not at all   (2) A little   (3) Not certain   (4) Quite a lot   (5) A lot

Thank you

Figure 2.24 Breakdown of response by the five respondents for the Anxiety Questionnaire survey.

Respondent 1	Respondent 2	Respondent 3	Respondent 4	Respondent 5
Q1. Male	Q1. Female	Q1. Female	Q1. Male	Q1. Male
Q2. 25	Q2. Aged 27	Q2. 33	Q2. 29	Q2. 21
Q3. A little anxious	Q3. Not at all	Q3. Anxious a lot	Q3. Not certain	Q3. Anxious a lot

Figure 2.25 Data that has been inputted from the Anxiety Questionnaire study.

Untitled - SPSS Data Editor

File Edit View Data Transform Analyze Graphs Utilities

2 : anxiety

	sex	age	anxiety	var
1	1.00	25.00	2.00	
2				
3				
4		.		

Exercise 5: Inputting data in SPSS for Windows

Now continue this for each respondent. The completed dataset is printed at the end of this chapter if you want to check it (after you have had a go!).

Saving data in SPSS for Windows

When you have finished inputting your data into SPSS successfully, you will want to save it. Whether you save it onto your hard drive or to a floppy drive will depend on whether you are at home or at college. However, it is the same as saving a Word file. Bring down the File menu and select the **Save As** option. A window entitled 'Newdata: Save Data As', like that in Figure 2.26, will open up. Choose where you want to save it, be it on a floppy disk or hard drive, by exploring the **Save in:** box.

When you have determined where you would like to save the file, then type in the **File Name:** (here, **anxiety**) and then press the **Save** button. Ensure that as you work and change a file, you keep saving it. As you know, computers can crash and if you are putting in a lot of data, nothing is so disheartening than to lose your work. So, as with all computers, constantly save your file. And if necessary make a back up.

Figure 2.26   Save Data As window.

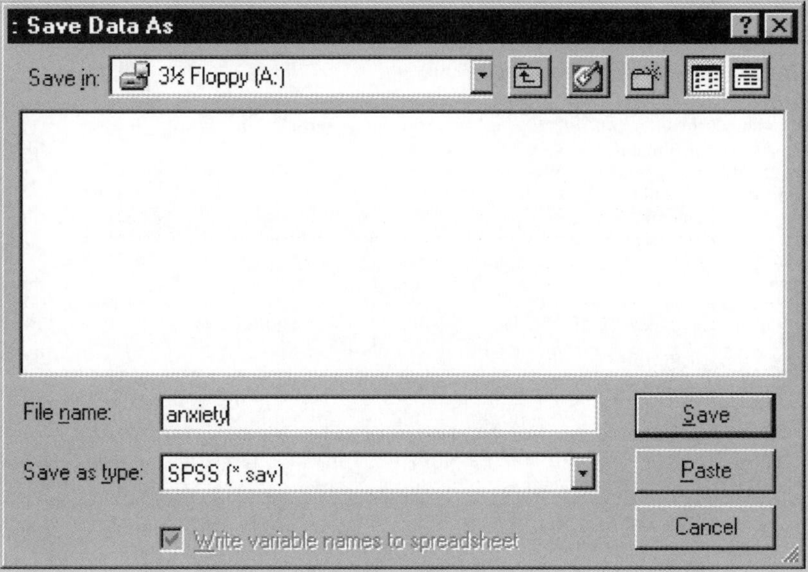

Introduction to the book's datafile

Throughout the rest of the book we will be using some data to demonstrate various statistical procedures. This datafile is on the Web at www.booksites. net/maltby and you will need to download it from the site for completing this section. Load up SPSS for Windows, pull down the **File** menu, and then press **Open**, and then **Data**. Select 3¹/₂ Floppy [A:] from **Look in** box and select Datafile.sav, and press **Open**. In this section we are going to introduce you to this data.

We have collected data from 60 respondents regarding their attitudes and behaviours on a number of variables. You will use all of these variables as you go through the book. We have deliberately kept the number of respondents and variables small so you don't become overwhelmed. What we are going to do here is introduce you to each variable in turn so you come to know the data a little, and it won't be completely new to you next time you come across the variable. When example items are given, try answering them yourself.

Variables 1–7 (named Vol1, Vol2, Vol3, Vol4, Vol5, Vol6 and Vol7)

These variables are all the items of a scale that is a measure of why people do, or would do, voluntary work. These are all shown in Figure 2.27. There are a number of altruistic (generous, socially interested) reasons, and a number of less altruistic reasons. Go on: try filling it in for yourself.

Figure 2.27   The Voluntary Work Questionnaire.

1. I (would) do voluntary work because it is a kind thing to do	Agree strongly	Agree	Not certain	Disagree	Disagree strongly
2. I (would) do voluntary work to help me in my career prospects	Agree strongly	Agree	Not certain	Disagree	Disagree strongly
3. I (would) do voluntary work to put something back into society	Agree strongly	Agree	Not certain	Disagree	Disagree strongly
4. My doing voluntary work will (would) look good on my Curriculum Vitae (CV)	Agree strongly	Agree	Not certain	Disagree	Disagree strongly
5. I (would) do voluntary work because it is good to be able to help the less fortunate	Agree strongly	Agree	Not certain	Disagree	Disagree strongly
6. I (would) do voluntary work because it impresses my friends and family	Agree strongly	Agree	Not certain	Disagree	Disagree strongly
7. I (would) do voluntary work because it performs an important function in society	Agree strongly	Agree	Not certain	Disagree	Disagree strongly

Variable 8 Sex (sex of the respondent)

Within this dataset males are coded as 1, females are coded as 2.

Variable 9: Age (age of the respondent)

Within this dataset age is coded as the number of years of the respondent.

Variable 10: BigI

Belief in Good Luck. This indicates attitudes that good luck exists and that fortunate things happen due to this good luck. The variable is derived from the Belief in Good Luck Scale (Darke and Freedman, 1997) which comprises 12 item measures. Of these, ten of the items concern the belief in good luck, and two concern belief in bad luck. Respondents are required to indicate the extent of their agreement using a Likert-type scale from *Strongly disagree* (1) through *Strongly agree* (6). The two items on bad luck are reverse scored. Example items include 'Luck plays an important part in everyone's life', 'Even the things in life I can't control tend to go my way because I'm lucky', and 'I consider myself to be a lucky person'. The scores you see in the datafile are total scores for each individual. Higher scores indicate a higher belief in good luck.

Variable 11: DeppHapp

A measure of Happiness which indicates good feelings, satisfaction, and positive feelings. The Depression–Happiness Scale (McGreal and Joseph, 1993) is a self-report questionnaire that contains 25 items measuring aspects of happiness and unhappiness (depression). Of these, 12 of the items concern positive thoughts and feelings, and 13 of the items concern negative thoughts and feelings. Example items include 'I felt sad', 'I felt cheerful', 'I felt happy', and 'I felt run down'. Respondents are asked about how they had felt in the past seven days, and asked to rate each item on a four-point scale: *Never* (0), *Rarely* (1), *Sometimes* (2), and *Often* (3). The 13 negative items are reverse scored so that possible scores on the scale can range between 0 and 75. Higher scores on the scale indicate a higher frequency of positive thoughts, feelings, and bodily experiences. The scores you see in the datafile are total scores for each individual. Higher scores indicate a higher level of happiness.

Variable 12: Neurot

A measure of the Personality variable Neuroticism which indicates feelings of anxiety, worry, moodiness and frequently depression. The scale is from the Abbreviated Form of the Revised Eysenck Personality Questionnaire (Francis *et al.*, 1992) and contains a six-item measure of neuroticism. Example items include 'Does your mood often go up and down?', 'Would you call yourself a nervous person?' and 'Do you suffer from nerves?'. Respondents are asked to rate each item on a two-point scale: Yes (scored as 1), and No (scored as 0). The scores you see in the datafile are total scores for each individual. Higher scores indicate a higher level of neuroticism.

Variable 13: Neurot2

A measure of the Personality variable Neuroticism measured six months after the administration of the original measure of Neuroticism (Variable 12). The scores you see in the datafile are total scores for each individual. Higher scores indicate a higher level of neuroticism.

Variable 14: Neurot3

A measure of the Personality variable Neuroticism measured one year after the administration of the original measure of Neuroticism (Variable 12). The scores you see in the datafile are total scores for each individual. Higher scores indicate a higher level of neuroticism.

Variable 15: Obsym

A measure of obsessional symptoms derived from the Sandler–Hazari Obsessionality Inventory (Sandler and Hazari, 1960). This inventory contains a measure of obsessional symptoms. Obsessional symptoms are with items describing feelings of guilt, ritualistic behaviours, indecision and compulsive thoughts and impulses. Example items include 'I often have to check up whether I have closed a door or switched off a light', 'I am often inwardly compelled to do certain things even though my reason tells me it is not necessary', and 'At times I feel a compulsion to count things'. Respondents are asked to rate each item on a two-point scale: Yes (scored as 1), and No (scored as 0). The scores you see in the datafile are total scores for each individual. Higher scores indicate a higher level of obsessional symptoms.

Variable 16: Reltype

Identification of the religious type for a person (Allport and Ross, 1967; Gorsuch and Venable, 1983). This is a categorical variable that indicates whether people are:

- extrinsically religious (socially religious; going to church because they enjoy seeing people they know and it helps them make friends);

- intrinsically religious (personally religious; important to have time in private thought and prayer, trying hard to live according to religious beliefs);

- not religious at all.

In the present sample, an Extrinsic religious person is scored as '3', an Intrinsic religious person is scored as '2', and a Non-religious person is scored as '1'.

Variable 17: Freqpow

Measurement of frequency of attendance at a place of worship. Respondents are asked to indicate how often they attend a place of worship. Possible responses are: 1 = Never, 2 = Rarely, 3 = Monthly, 4 = Weekly, 5 = Once a week or more. The higher the score on this variable, the more frequently individuals attend a place of worship.

Variable 18: Freqpp

Measurement of frequency of personal prayer. Respondents are asked to indicate how often they indulge in personal prayer. Possible responses are: 1 = Never, 2 = Rarely, 3 = Sometimes, 4 = Weekly, 5 = Daily. The higher the score on this variable, the more frequently individuals practice personal prayer.

Variable 19: Religru

A measure of how religious respondents view themselves. With this question respondents are asked 'How religious are you?' Responses were scored on a five-point scale: 1 = Not a lot, 2 = A little, 3 = Slightly, 4 = Quite a lot, 5 = Very much so. The higher the score on this variable, the more respondents view themselves as religious.

Variable 20: Freqpow2

Measurement of frequency of attendance at a place of worship six months after the original scale had been administered (measured by Freqpow; Variable 17). Respondents are asked to indicate how often they attend a place of worship. Possible responses are: 1 = Never, 2 = Rarely, 3 = Monthly, 4 = Weekly, 5 = Once a week or more. The higher the score on this variable, the more frequently individuals attend a place of worship.

Variable 21: Freqpp2

Measurement frequency of personal prayer six months after the original scale had been administered (measured by Freqpp; Variable 18). Respondents are asked to indicate how often they indulge in personal prayer. Possible responses are: 1 = Never, 2 = Rarely, 3 = Sometimes, 4 = Weekly, 5 = Daily. The higher the score on this variable, the more frequently individuals practice personal prayer.

Variable 22: Freqpp3

Measurement frequency of personal prayer one year after the original scale had been administered (measured by Freqpp; Variable 18). Respondents are asked to indicate how often they indulge in personal prayer. Possible responses are: 1 = Never, 2 = Rarely, 3 = Sometimes, 4 = Weekly, 5 = Daily. The higher the score on this variable, the more frequently individuals practice personal prayer.

Summary of variables

A full summary of the variables is given in Figure 2.28.

Figure 2.28   Variable summary table.

	Variable name	Variable label	Values
1	Vol1	Item 1 of the Voluntary Work Scale: 'I (would) do voluntary work because it is a kind thing to do'	(5) Agree strongly, (4) Agree, (3) Uncertain, (2) Disagree, (1) Disagree strongly
2	Vol2	Item 2 of the Voluntary Work Scale: 'I (would) do voluntary work to help me in my career prospects'	(5) Agree strongly, (4) Agree, (3) Uncertain, (2) Disagree, (1) Disagree strongly
3	Vol3	Item 3 of the Voluntary Work Scale: 'I (would) do voluntary work to put something back into society'	(5) Agree strongly, (4) Agree, (3) Uncertain, (2) Disagree, (1) Disagree strongly
4	Vol4	Item 4 of the Voluntary Work Scale: 'My doing voluntary work will (would) look good on my Curriculum Vitae (CV)'	(5) Agree strongly, (4) Agree, (3) Uncertain, (2) Disagree, (1) Disagree strongly
5	Vol5	Item 5 of the Voluntary Work Scale: 'I (would) do voluntary work because it is good to be able help the less fortunate'	(5) Agree strongly, (4) Agree, (3) Uncertain, (2) Disagree, (1) Disagree strongly
6	Vol6	Item 6 of the Voluntary Work Scale: 'I (would) do voluntary work because it impresses my friends and family'	(5) Agree strongly, (4) Agree, (3) Uncertain, (2) Disagree, (1) Disagree strongly.
7	Vol7	Item 7 of the Voluntary Work Scale: 'I (would) do voluntary work because it performs an important function in society'	(5) Agree strongly, (4) Agree (3) Uncertain, (2) Disagree, (1) Disagree strongly

continued

Figure 2.28 (*continued*)

	Variable name	Variable label	Values
8	Sex	Sex of the respondent	1 = male, 2 = female
9	Age	Age of the respondent	Number equals the age of the respondent
10	Bigl	Belief in Good Luck which indicates attitudes that good luck exists and that fortunate things that happen are due to this good luck	Higher scores indicate a higher belief in good luck
11	Depphapp	A measure of Happiness which indicates good feelings, satisfaction, positive feelings	Higher scores indicate a higher level of happiness
12	Neurot	A measure of the Personality variable Neuroticism which indicates feelings of worry, nervousness and anxiety	Higher scores indicate a higher level of neuroticism
13	Neurot2	A measure of the Personality variable Neuroticism measured six months after the administration of the Neurot	Higher scores indicate a higher level of neuroticism
14	Neurot3	A measure of the Personality variable Neuroticism measured one year after the administration of the Neurot	Higher scores indicate a higher level of neuroticism
15	Obsym	A measure of Obsessional Symptoms that indicates obsessive behaviours such as feelings of guilt, worry and obsessions	Higher scores indicate a higher level of obsessional symptoms
16	Reltype	Identification of religious type of the person. A categorical variable that indicates whether people are either extrinsically religious (socially religious), intrinsically religious (personally religious), or not religious at all	3 = Extrinsic, 2 = Intrinsic, 1 = Not religious
17	Freqpow	Measurement of frequency of attendance at a place of worship	1 = Never, 2 = Rarely, 3 = Monthly, 4 = Weekly, 5 = Once a week or more
18	Freqpp	Measurement of frequency of personal prayer	1 = Never, 2 = Rarely, 3 = Sometimes, 4 = Weekly, 5 = Daily
19	Religru	Measurement of self-rating of religiosity	1 = Not a lot, 2 = A little, 3 = Slightly, 4 = Quite a lot, 5 = Very much so
20	Freqpow2	Measurement of frequency of attendance at a place of worship administered six months after the first administration of Freqpow	1 = Never, 2 = Rarely, 3 = Monthly, 4 = Weekly, 5 = Once a week or more
21	Freqpp2	Measurement of frequency of personal prayer administered six months after the first administration of Freqpow	1 = Never, 2 = Rarely, 3 = Sometimes, 4 = Weekly, 5 = Daily
22	Freqpp2	Measurement of frequency of personal prayer administered one year after the first administration of Freqpow	1 = Never, 2 = Rarely, 3 = Sometimes, 4 = Weekly, 5 = Daily

References

Allport, G.W. and Ross, J.M. (1967). Personal religious orientation and prejudice. *Journal of Personality and Social Psychology*, **5**, 432–433.

Barr, W. (2000). Characteristics of severely mentally ill patients in and out of contact with community mental health services. *Journal of Advanced Nursing*, **31**, 1189–1198.

Brown, S. (2000). Overfishing and global warming land cod on endangered list. *The Guardian*, 20 July.

Brown, S.A., Harrist, R.B., Villagomez, E.T., Segura, M., Barton, S.A. and Hanis, C.L. (2000). Gender and treatment differences in knowledge, health beliefs, and metabolic control in Mexican Americans with type B diabetes. *Diabetes Educator*, **26**, 425–428.

Chrisafis, A. (2000). Conveyancing website aims to cut gazumping and stress of moving home. *The Guardian*, 20 July.

Darke, P.R. and Freedman, J.L. (1997). The Belief in Good Luck Scale. *Journal of Research in Personality*, **31**, 486–511.

Denny, C. (2000). Big spend may bring [inflation] rate rise, says MPC. *The Guardian*, 20 July.

Francis, L.J., Brown, L.B. and Philipchalk, R. (1992). The development of an Abbreviated Form of the Revised Eysenck Personality Questionnaire (Epqr-A) – its use among students in England, Canada, the USA and Australia. *Personality and Individual Differences*, **13**, 443–449.

Gorsuch, R.L. and Venable, G.D. (1983). Development of an 'Age Universal' I-E scale. *Journal for the Scientific Study of Religion*, **22**, 181–187.

Kettle, M. (2000). US elections prove a TV turn-off. *The Guardian*, 21 July.

McGreal, R. and Joseph, S. (1993). The Depression–Happiness Scale. *Psychological Reports*, **73**, 1279–1282.

Meikle, J. (2000). Scientists warn of 30% rise in human BSE. *The Guardian*, 18 July.

Naerde, A., Tambs, K., Mathiesen, K.S., Dalgard, O.S. and Samuelsen, S.O. (2000). Symptoms of anxiety and depression among mothers of pre-school children: effect of chronic strain related to children and child care-taking. *Journal of Affective Disorders*, **58**, 181–199.

Sandler, J. and Hazari, A. (1960). The obsessional: on the psychological classification of obsessional character traits and symptoms. *British Journal of Medical Psychology*, **33**, 113–122.

Scott, K. (2000). Miners' long hours blamed as lung disease returns. *The Guardian*, 18 July.

Answers to exercises

Exercise 1: Chapter energiser

See Figure 2.29 for the solution to the Number-cross Puzzle.

Figure 2.29   Solution to Number-cross Puzzle.

1 2	2	2 9	■	3 2	■	4 6	7	5 8
7	■	6 4	2	0	0	2	■	8
7 3	8 9	8	■	5	■	9 2	10 9	7
■	2	■	■	7	■	■	9	■
11 7	7	8	5	■	12 9	8	7	0
■	1	■	■	13 6	■	■	0	■
14 8	4	15 1	■	9	■	16 6	4	17 4
7	■	18 2	6	7	3	9	■	8
19 4	8	4	■	7	■	20 2	5	2

Exercise 2: Identifying variables

From Newspaper article headlines

1. 'Overfishing and global warming land cod on endangered list': (1) level of fishing; (2) level of global warming; and (3) levels of cod.

2. 'Big spend may bring [inflation] rate rise, says MPC': (1) level of spending; and (2) levels of inflation.

3. 'Conveyancing website aims to cut gazumping and stress of moving home': (1) amount of gazumping; and (2) level of stress.

From Journal article headlines

1. 'Symptoms of anxiety and depression among mothers of pre-school children: effect of chronic strain related to children and child care-taking': (1) level of anxiety; (2) level of depression; and (3) level of chronic strain.

2. 'Characteristics of severely mentally ill patients in and out of contact with community mental health services': (1) characteristics of mentally ill patients; and (2) whether the patient is in contact with community mental health services.

3. 'Gender and treatment differences in knowledge, health beliefs, and metabolic control in Mexican Americans with type B diabetes': (1) gender; (2) treatment type; (3) health beliefs; and (4) level of metabolic control.

Exercise 3: Identifying more variables

We suggest the following variables can be found in the article 'US elections prove a TV turn-off'. We have little doubt that you will have found other ones.

- Rating of 'points' (percentage of people watching). This is a continuous variable.
- Whether voters intended to watch any of the Republican convention. This is a categorical variable.
- The percentage of voters intendinged to watch any of the Republican convention. This is a continuous variable.
- Whether voters intended to watch any of the Democrat convention. This is a categorical variable.
- The percentage of voters intending to watch any of the Democrat convention. This is a continuous variable.

Exercise 4: Using a dataset

1. Respondent 2 has blue eyes.
2. Respondent 2 is 35.
3. There are three females in the present sample.
4. There are three people over 30 in the present sample.

Exercise 5: Inputting data in SPSS for Windows

Your dataset from the Anxiety Questionnaire should look like that in Figure 2.30 when completed.

Figure 2.30   How your datafile from the Anxiety Questionnaire should look.

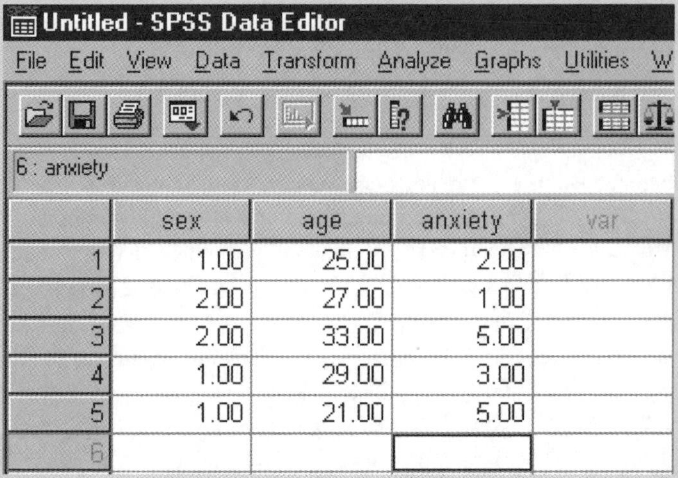

CHAPTER 3

Changing, combining and describing variables

Mathematics is *a language.*

(Josiah Willard Gibbs, 1839–1903)

In this chapter we shall:

- introduce you to multi-question variables and to terms such as 'recode' and 'compute', used in SPSS for Windows;
- introduce you to descriptive statistics, and more specifically frequency counts, averages, a measure of dispersion, bar-charts, and histograms.

In this chapter, aspects of SPSS for Windows will be illustrated using the dataset 'Datafile.sav', so ensure you load up this dataset when using SPSS for Windows.

Exercise 1: Chapter energiser

Try the simple exercises given in Figure 3.1. We are using these exercises as they cover many of the skills and procedures explained in this chapter, and are a good way of putting you in the right frame of mind for tackling these skills.

Changing and combining variables (recoding and computing variables)

So far, we have considered single variables. However, there are many occasions when variables are either changed, or can be used together (the latter being most common when using continuous variables). In this chapter, we are going to show you how researchers change and combine variables, to produce new variables.

Figure 3.1   Chapter energiser.

Question 1: For each set of numbers, add the numbers together.

1. 4, 3, 2, 1, 5
2. 3, 2, 1, 2, 5

Question 2: For each set of numbers, rank the numbers in ascending (lowest to highest) order.

1. 7, 5, 2, 4, 3, 9, 17
2. 2, 7, 62, 53, 1, 2

Question 3: For each set of numbers, rank the numbers in descending (highest to lowest) order.

1. 15, 5, 12, 4, 31, 9, 17
2. 23, 7, 2, 53, 11, 2

Question 4: For each set of numbers, rank the numbers in ascending order. What is the middle number?

1. 1, 9, 5, 4, 3, 2, 8
2. 19, 13, 17, 18, 12, 21, 23

Question 5: For each set of numbers, what is the most popular number?

1. 2, 56, 3, 3, 2, 2, 13, 23
2. 117, 112, 113, 117, 119

Question 6: For each set of numbers, (1) add together the numbers in each set, and (2) count how many numbers in each set. Then divide your finding for (1), by your finding for (2).

1. 1, 7, 6, 3, 3, 4
2. 23, 27, 18, 12, 20

To illustrate how we change and combine variables, we are going to use an example of an existing standard measure (Figure 3.2). Read the description of the Depression–Happiness Scale (McGreal and Joseph, 1993). Don't worry if you are unsure of any of the terminology used in the description: the chapter is designed to explore what some of these terms mean. Within the scale description, there are two points to consider: (i) that some of the items are 'reversed'; and (ii) that all the items are used together to measure depression–happiness, with higher scores indicating a higher level of happiness (positive thoughts, feelings, and bodily experiences). We will use these two points to explore the two main ways that research statisticians (i) change and (ii) combine data.

Changing (reversing) data

The reviewer of the Depression–Happiness Scale lists a number of variables that are designed to measure happiness. Let us consider only four of these variables to understand how researcher statisticians change data:

Figure 3.2   Description of the Depression–Happiness Scale.

'The Depression–Happiness Scale (McGreal and Joseph, 1993) is a self-report questionnaire that contains 25 items measuring aspects of happiness and unhappiness (depression). Of these, 12 of the items concern positive thoughts and feelings, and 13 of the items concern negative thoughts and feelings. Respondents are asked about how they have felt in the preceeding seven days, and asked to rate each item on a four-point scale: *Never* (0), *Rarely* (1), *Sometimes* (2), and *Often* (3). The 13 negative items are reverse scored so that possible scores on the scale can range between 0 and 75. Higher scores on the scale indicate a higher frequency of positive thoughts, feelings, and bodily experiences'.

(Lewis, 2000, p. 568)

Scale items and instructions

Read each statement and circle the number that best describes how frequently each statement was true for you in the past seven days. You may have experienced both positive and negative feelings at different times.

0 = Never 1 = Rarely					2 = Sometimes 3 = Often				
1. I felt sad	0	1	2	3	14. I felt unattractive	0	1	2	3
2. I felt I had failed as a person	0	1	2	3	15. I felt optimistic about my future	0	1	2	3
3. I felt dissatisfied with my life	0	1	2	3	16. I felt life was rewarding	0	1	2	3
4. I felt mentally alert	0	1	2	3	17. I felt cheerless	0	1	2	3
5. I felt disappointed with myself	0	1	2	3	18. I felt life had a purpose	0	1	2	3
6. I felt cheerful	0	1	2	3	19. I felt too tired to do anything	0	1	2	3
7. I felt life wasn't worth living	0	1	2	3	20. I felt pleased with the way I am	0	1	2	3
8. I felt satisfied with my life	0	1	2	3	21. I felt lethargic	0	1	2	3
9. I felt healthy	0	1	2	3	22. I found it was easy to make decisions	0	1	2	3
10. I felt like crying	0	1	2	3	23. I felt life was enjoyable	0	1	2	3
11. I felt I had been successful	0	1	2	3	24. I felt life was meaningless	0	1	2	3
12. I felt happy	0	1	2	3	25. I felt run down	0	1	2	3
13. I felt I couldn't make decisions	0	1	2	3					

(Reproduced with permission of authors and publisher from: McGreal, R., & Joseph, S. (1993). The Depression–Happiness Scale. *Psychological Reports*, 1993, 73, 1279–1282 © *Psychological Reports*, 1993.)

- I felt sad (variable 1).
- I felt cheerful (variable 6).
- I felt happy (variable 12).
- I felt run down (variable 25).

Respondents are asked to rate their feelings on a four-point scale: (0) Never, (1) Rarely, (2) Sometimes, and (3) Often. Therefore, a person who is *happy* may respond to the questions as in Figure 3.3(a) and an *unhappy or depressed* person may respond as in Figure 3.3(b).

However, to use these responses for statistical purposes we need to note that two of the variables are known as 'positive' items and two are known as 'negative' items:

- I felt sad (negative).
- I felt cheerful (positive).
- I felt happy (positive).
- I felt run down (negative).

Lewis (2000) notes that higher scores on the Depression–Happiness Scale indicate a higher level of happiness. At the moment, the items 'I felt cheerful' and 'I felt happy' are consistent with the coding for higher scores showing happiness, because if respondents reported feeling either of these 'Often' they would score the highest score, '3', and if they 'Never' felt happy or cheerful they would get the lowest score, '0'. However, this is not the case for the other two items, 'I felt sad' and 'I felt run down', as getting higher scores for these variables would indicate a higher level of *un*happiness.

Therefore, when scoring responses to the scale, we need to change the scoring given for these items. We do this by assigning '3' instead of '0', '2' instead of '1', '1' instead of '2' and '0' instead of '3' to responses for these items. This means that now the happy person would get a score of:

- 3 for 'Never' feeling sad (reversed item);
- 3 for 'Never' feeling run down (reversed item);
- 3 for 'Often' feeling happy;
- 3 for 'Often' feeling cheerful.

Figure 3.3   Example answers to some of the variables of the Depression–Happiness Scale.

0 = Never		1 = Rarely			2 = Sometimes		3 = Often	
Happy person					Sad person			
I felt sad	0✔	1	2	3	I felt sad	0	1	2	3✔
I felt cheerful	0	1	2	3✔	I felt cheerful	0✔	1	2	3
I felt happy	0	1	2	3✔	I felt happy	0✔	1	2	3
I felt run down	0✔	1	2	3	I felt run down	0	1	2	3✔
(a)					(b)			

The sad person would get a score of:

- 0 for 'Often' feeling sad (reversed item);
- 0 for 'Often' feeling run down (reversed item);
- 0 for 'Never' feeling happy;
- 0 for 'Never' feeling cheerful.

This whole process of changing variables in this way is known as 'reversing' items. By *reversing* the scores for some of the items on the Depression–Happiness Scale, the scores given for responses are now consistent with the way the reviewer described the scoring for this scale.

Combining (computing) scores

It is often the case that the variables cannot be measured by a single measure. There may be several elements within a single variable. Variables such as depression or happiness are made up of many different types of thoughts and feelings. This is why the Depression–Happiness Scale uses a number of positive and negative feelings and thoughts. This has been found to be the case for many variables. Therefore, researchers, rather than using single variables, tend to use a number of variables together to measure something.

This is fairly straightforward when you have already reversed items. In the description of the Depression–Happiness Scale the reviewer indicates that responses to items are combined to create an overall measure of happiness. That is, the scores are added together to produce an overall score. So, all you do is simply add the scores from the responses to each item to compute the overall scale score. Let us use the example for the happy and unhappy person in the example used on the previous page and which is summarised in Figure 3.4.

Figure 3.4   *Examples of reversed items.*

The happy person would get a score of:
- 3 for 'Never' feeling sad (reversed item)
- 3 for 'Never' feeling run down (reversed item), as well as
- 3 for 'Often' feeling happy, and
- 3 for 'Often' feeling cheerful

The *un*happy person would get a score of:
- 0 for 'Often' feeling sad (reversed item)
- 0 for 'Often' feeling run down (reversed item), as well as
- 0 for 'Never' feeling happy, and
- 0 for 'Never' feeling cheerful

So, using the example of reversed items the total scores for the Depression–Happiness Scale would be as follows:

- The happy persons gets '3' for each response to the four variables (after reversing the items), so $3 + 3 + 3 + 3 = 12$. Therefore our happy person scores '12' overall on the Depression–Happiness Scale.

- The unhappy person gets '0' for each response (after reversing the items), so $0 + 0 + 0 + 0 = 0$, meaning our unhappy person scores '0' overall on the Depression–Happiness Scale.

As Lewis (2000) notes, higher scores on the Depression–Happiness Scale indicate a higher level of happiness. Our findings among our two respondents confirm this, as the happy person scores higher than the unhappy person.

The whole process of adding variables in this way is known as 'computing' items.

Exercise 2: Reversing and computing scores on a scale

Consider the description of the Francis Scale of Attitude towards Christianity (Figure 3.5), and the four example items.

1. If, on this scale, higher scores indicate a more positive attitude towards Christianity, which two items need to be reversed before computing overall scores for the scale?

2. If a respondent answered 'Agree' to item 1, 'Disagree strongly' to item 2, 'Strongly agree' to item 3, and 'Disagree' to item 4, what would the overall scores be, after reversing the scores for the appropriate items?

Figure 3.5   Description of the Francis Scale of Attitude towards Christianity.

The Francis Scale of Attitude towards Christianity (Francis and Stubbs, 1987) is a 24-item scale concerned with responses to statements about God, Jesus, Bible, prayer, and church, scored on a five-point scale ranging from (5) *Agree strongly* through (3) *Uncertain* to (1) *Disagree strongly*. Possible scores range from 24 to 120, with higher scores indicating a more positive attitude towards Christianity.

Example items and response format:

1. God means a lot to me
2. I think going to church is a waste of my time
3. Prayer helps me a lot
4. I think the Bible is out of date

(5) *Agree strongly*, (4) *Agree* (3) *Uncertain*, (2) *Disagree* (1) *Disagree strongly*

Reversing and computing items on SPSS for Windows

We are now going to show you how to reverse and compute items using SPSS for Windows. To do this you will need to load up the dataset 'datafile.sav' introduced to you in Chapter 2, and provided on the disk.

For these examples we will be using seven variables listed in the dataset, Vol1 to Vol7 (Figure 3.6). These variables are used together to measure individual's reasons for doing voluntary work. In the dataset for SPSS for Windows the variables Vol1 to Vol7 represent each of the scale items (Vol1 = item 1, Vol2 = item 2, etc.), with scores for responses to each of the variables for 60 respondents. We are now, in SPSS for Windows, going to compute overall scores on the Voluntary Work Scale for each respondent. However, a number of items need to be reversed before scoring. Therefore, we are going to do two things: (i) reverse scores in SPSS for Windows; and then (ii) compute scores in SPSS for Windows.

Figure 3.6   Description of the Voluntary Work Scale.

The Voluntary Work Scale is a seven-item scale concerned with measuring the reasons why people undertake (or may undertake) voluntary work. Four of the items are designed to measure positive, altruistic reasons for doing voluntary work, while three of the items are designed to indicate selfish reasons for doing voluntary work. Response to these three items should be reversed before computing overall scores for the scale. Responses are scored on a five-point scale: (5) *Agree strongly*, (4) *Agree*, (3) *Uncertain*, (2) *Disagree*, (1) *Disagree strongly*. Possible scores range from 7 to 35, and higher overall scores should indicate more altruistic reasons for doing voluntary work.

1. I (would) do voluntary work because it is a kind thing to do
2. I (would) do voluntary work to help me in my career prospects (R)
3. I (would) do voluntary work to put something back into society
4. My doing voluntary work will (would) look good on my Curriculum Vitae (CV) (R)
5. I (would) do voluntary work because it is good to be able help the less fortunate
6. I (would) do voluntary work because it impresses my friends and family (R)
7. I (would) do voluntary work because it performs an important function in society

(R) = reversed item

Reversing coding in SPSS for Windows

For the Voluntary Work Scale there are four positive items and three negative items.

Positive worded items

- Vol1 (kind thing)
- Vol3 (something back to society)
- Vol5 (help the less fortunate)
- Vol7 (important function in society)

Negative worded items

- Vol2 (career prospects)
- Vol4 (Curriculum Vitae)
- Vol6 (impresses my friends and family)

You need to recode Vol2, Vol4, and Vol6. Pull down the **Transform** menu, click on **Recode**. Another menu should appear, click on **Into Same Variables**. You should get the box **Recode into Same Variables** as in Figure 3.7. Transfer Vol2, Vol4 and Vol6 from the box on the left, into the box **Numeric Variables:** using the > button.

Now click on the **Old and New Values** button. You will get the submenu **Recode into Same Variables: Old and New Values** as in Figure 3.8.

Now you need to tell the computer what values to recode. Here, there are five possible answers to each question. For variables Vol2, Vol4, and Vol6, answers scored as '5' need to be scored as '1', answers scored as '4' need to be scored as '2', etc. To do this, type '5' in **Value** box on the left hand side (the **Old Value** side), and type '1' in the **Value** box on the right hand side.

Figure 3.7   Recode into Same Variables.

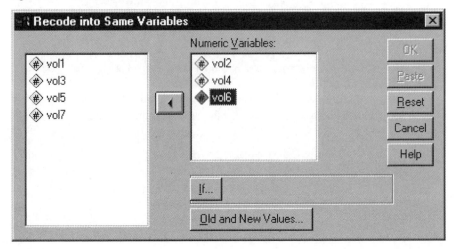

Figure 3.8   Recode into Same Variables: Old and New Values.

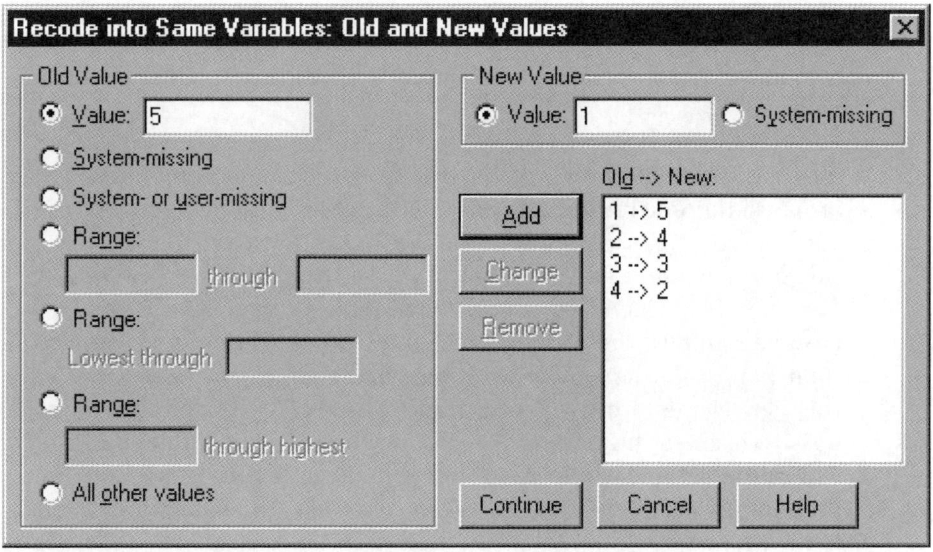

Then press the **Add** button next to the **Old –> New:** box. Then do the same for the rest of the recoding.

* Type '4' in **Value** box on the left hand side (the **Old Value** side), and type '2' in the **Value** box on the right hand side, and then press the **Add** button next to the **Old –> New:** box.

* Type '3' in **Value** box on the left hand side (the **Old Value** side), and type '3' in the **Value** box on the right hand side, and then press the **Add** button next to the **Old –> New:** box. And so on.

When you have included the coding for recoding '2' values into '4' values, and '1' into '5' values, pressing **Add** each time, press **Continue**. You will now go back to the **Recode into Same Variables** box. Press **OK**. The variables Vol2, Vol4 and Vol6 are now recoded.

Recoding into different variables

Using the 'Recode into Different Variables' procedure, you are able to create a new variable that alters the code given to values, but leaves the original variable as it is in the dataset (for example, you want to use the original variable with its original codes at a later date). One of the most common examples of recoding like this is when you want to condense the levels of a variable, such as categorising age of respondents. For example, you might want to examine differences in the samples between young, middle aged and elderly people. This might involve putting all individuals under 25 into a 'young persons' group (coded as 1), all individuals aged 40–50 years into a 'middle aged' group (coded as 2), and all individuals aged over 60 into a 'elderly' group (coded as 3).

To do this in SPSS for Windows, click on the **Transform** pull-down menu, click on **Recode**, and then **Into Different variable**. You will arrive at the **Recode into Different Variables** window. Transfer the variable you wish to recode from the Variable list on the left hand side of the window and transfer it into the **Input Variable –> Output Variable:** box. Then you need to give the new variable you wish to create a name in the **Name** slot in the **Output Variable** box on the right hand side (remember no more than eight characters). Transfer this new Name into the centre **Input Variable –> Output Variable:** box. To reassign the values, click on the **Old and New Values** and you will go to the same window as you had for **Recode into Same Variable** procedure. Recode the variable values by placing old values in the left hand box, and new values in the right hand box. Remember to press **Add** for each recode. When you have finished press **Continue** and then **OK** and SPSS for Windows will compute the new values. You will be able to check this procedure by seeing if new variables are created in the dataset.

Try this skill with the Age variable. Create a new variable **AgeCat** (Age Category) and place all individuals under 25 into a 'young persons' group (coded as 1), all individuals aged 40–50 years into a 'middle aged' group (coded as 2), and all individuals aged over 60 into a 'elderly' group (coded as 3).

Computing in SPSS for Windows

We now need to compute total scores for each individual on the Voluntary Work Scale. Pull down the the **Transform** menu. Click on **Compute**. You will get the **Compute Variable** menu (Figure 3.9). Type **VolScale** (though this can be anything under eight letters) into the **Target Variable:** box. Transfer vol1 from the box on the left into the **Numeric Expression** box. Now click on the + button on the numeric keypad section just below that. Transfer vol2 from the box on the left into the **Numeric Expression** box and click on the + button on the numeric keypad section just below that. Repeat this for vol3, vol4, vol5, vol6 and vol7, pressing + each time, except after vol7. When you have done that press **OK**. You will return to the data sheet, and, if you explore along the data, you will see the new variable has been created at the end.

This is how you 'recode' and 'compute' variables in SPSS for Windows.

Figure 3.9   Compute Variable.

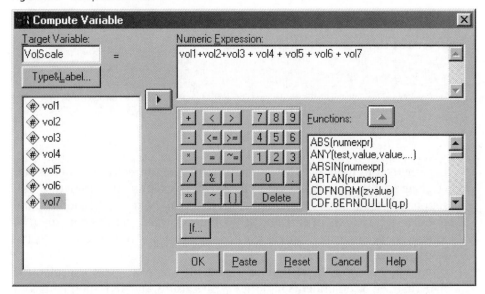

Describing variables (descriptive statistics)

Descriptive statistics are ways of describing data. There are many different types of descriptive statistics, but the aim of the book is to teach the most used and relevant statistics. Using these criteria, teaching descriptive statistics becomes fairly directed, and you will find, in most journal articles, that these are the main statistics that researchers use. The two main areas of descriptive statistics that we will cover are:

1.   Frequencies, averages and a measure of dispersion.
2.   Charts which can be used to examine data.

Frequencies, averages and a measure of dispersion

The two aspects of statistics that we will cover in this area represent the ways that we can use numerical information to describe our data.

Frequencies

The first aspect is frequencies of data. This information can be used to break down any variable, and to tell the researcher how many respondents answered at each level of the variable. Consider an example where a researcher asked 100 local residents whether they agree with the planned

Table 3.1   Question: Do you agree with the planned hypermarket in this area?

Possible answers	Frequency of answers
Agree	47 respondents answered the question with this response
Unsure	5 respondents answered the question with this response
Disagree	48 respondents answered the question with this response

hypermarket in their area. Respondents were given three choices of answer: (1) 'I agree', (2) 'Unsure', and (3) 'I disagree'. To examine the strength of the support for the planned hypermarket in the area, the researcher adds up the number of responses to each of the possible answers, and presents them in a table so he/she can examine the frequency of answers to each possible response (see Table 3.1).

On this occasion 47 respondents said they 'agreed' with the planned hypermarket, 5 respondents said they were 'unsure' and 48 respondents said they 'disagreed'. As we can see opinion is split between those respondents who agree with the planned hypermarket and those who disagree, with some local residents unsure of their position. This type of breakdown of answers is how frequencies are used. Frequencies are most often presented in the form of a table (Table 3.1).

Averages

Consider the extracts from two news articles (Figure 3.10) by Julia Hartley-Brewer (1999) and Marilyn Wann (1999) that appeared in *The Guardian*.

You have probably heard/read the phrase 'the average person thinks . . .'. Often when this phrase is used the writer is trying to indicate 'Most people think . . .'. Averages are ways in which researchers can summarise frequency data and find out what are the most common responses. Both these articles use averages to describe a situation or event. Article 1 uses averages around spending to show how Britons are spending their leisure time, and Article 2 uses the average women's dress size to examine debates around the effect of fashion.

However, there are three ways that researchers can use averages. These are known as mean, median and mode.

- *Mean*: this is calculated by adding all the values from each response to the variable, divided by the number of respondents.
- *Median*: this is calculated by putting all the values from responses to the variable, in order, from the smallest value to the largest value, and selecting the value that appears in the middle.
- *Mode*: this is the value that occurs most often in the data.

Figure 3.10   'Fun-loving Britons splash out most on leisure' and 'Self-hatred and celery sticks'.

Article 1

Fun-loving Britons splash out most on leisure: For the first time, the average family now spends more on holidays, home computers, and sport than on food and housing

Julia Hartley-Brewer
Thursday, 25 November 1999

Britons now spend more on fun than on any other household expenditure, as spending on leisure outstrips that on housing and food for the average family for the first time. Spending on holidays, home computers, sporting events and other leisure goods and services now forms the largest share of household outgoings, according to official figures published today. Until now, spending on food and non-alcoholic drinks has taken the largest slice of the average household's budget. The average household now spends a total of £350 a week, with £60 going on leisure goods and services, £59 on food and £57 on housing costs such as rents or mortgages, the government's annual survey of household spending reveals. A year ago, the same household spent £329 a week, with £55 spent on leisure, £56 on food and £51 on housing.

Article 2

Self-hatred and celery sticks: Life is too short for either. Marilyn Wann – size 28 and proud of it – explains why she is against Spain's decision to ban skinnies from the catwalk

Marilyn Wann
Monday, 29 March 1999

What size do you wear? An eight? A 23 and a half? A 40 extra long? Do I hear 60? And what do these numbers really say about you? Evidently, if you're a model and you're smaller than a British size 10, you're no longer welcome at The Salon Gaudi, Spain's annual fashion show. Salon director Paco Flaque said, 'If we promote the image of skinny women we are hurting our young people, and I am against that.'

(Reproduced with permission of authors and publisher from Hartley-Brewer (1999) and Wann (1999). © *The Guardian*, 1999.)

So consider this example. A researcher has asked five households how many children are living in the house. The researcher collates the information into frequencies, and displays the data in the form of a table (Table 3.2). The researcher wishes to work out the mean, median and mode.

For the *mean* the researcher adds all the responses, 0 + 2 + 2 + 3 + 4 (one household had no children, two households had 2 children, one household had 3 children, and one household had 4 children) and divides the total 11 (0 + 2 + 2 + 3 + 4 = 11) by the number of respondents asked. Here, five households took part in the research, so 11 is divided by 5. Therefore, the mean equals (=) 11 divided by 5, which is 2.2. Therefore, there is a mean of 2.2 children living in each household.

Table 3.2   Question: How many children in the household?

Possible answers	Frequency of answers
0	1 household
1	
2	2 households
3	1 household
4	1 household
5	
6	
7	
Continued . . .	

For the *median*, rank the order of numbers from the smallest to the highest; so, 0, 2, 2, 3, 4 and then select the middle number, which is 2. Therefore, there is a median of 2 children living within the household. This is fairly straightforward when the researcher has an odd number of cases. When there is an even number, you select the middle two numbers and divide them by two. So in the case of 0, 2, 2, 3, 4, 4, you would add together 2 and 3 and divide by 2; 5 divided by 2 = 2.5.

For the *mode*, the most common number is selected. In the example 0, 2, 2, 3, 4, number 2 is the mode. Therefore, there is a mode of 2 children living within each household.

Therefore, the researcher will report that among his sample the mean average of children in the household is 2.2, the median is 2.5 and mode is 2.

However, it is not common practice to report all three measures of average. Rather, the mean is the most commonly used, and as a rule the median and the mode tend to be used only when researchers suspect that reporting the mean may not represent a fair summary of the data.

Exercise 3: Calculating the mean, median and mode

1. Work out the mean, median and mode for the following set of numbers: 1, 2, 3, 3, 15.
2. Compare the mean score with the median and mode averages. Do you know why the mean is different from the median and mode?

As mentioned, the mean average is the most often used average reported in research. You will often find the mean used in articles, similar to the way it is used in Article 1, 'Fun-loving Britons splash out most on leisure', reporting on average spending. However, the mean is vulnerable to occasions when there is a number that is very different to the rest of the numbers. In the example above the number 15 distorts the data. Therefore, the median or the mode might represent a better description of the data. Clothes size is a common example to illustrate the use of the mode. Imagine five women whose dress sizes were 10, 14, 14, 14, 14. The mean average of this sample is 13.2. If the shop took this into account and only sold the average, then they would sell nothing to any of the five ladies because either the dress would be too big or too small. However, if the shop stocked the mode average, 14, then they would sell four dresses.

It is of interest to use the right measurement. Researchers often use median to address the same problem. An example of this would be when reporting the average age of the sample. Often researchers report the mean (and standard deviation). However, sometimes researchers find, particularly when using university samples, that although the majority of the sample are from an 18–21-year-old age range, the sample also comprises students who are studying late in life. This would bring the mean up. Therefore, researchers would additionally report the median average, to show this possible distortion, and to indicate that the average age was much younger.

Standard deviation

The standard deviation is a descriptive statistic that accompanies the use of the mean. The standard deviation provides the researcher with an indicator of how scores for variables are spread around the mean average. Standard deviation scores are relative, but the general rule is that the higher the standard deviation, the more scores around the mean are spread out. Therefore, a higher standard deviation statistic would be found for this set of five scores, 0, 25, 66, 78, 100, than for this set of five scores, 1, 2, 2, 3, 4. It is good practice always to report the standard deviation, when reporting the mean.

We are now going to develop your skills with statistics by showing you how to work out the standard deviation by hand. There are two formulae for working out the standard deviation. However, because you are a researcher, we are going to use the one most commonly used. This is the standard deviation used with data from samples.

A researcher has decided to examine how many cans of Super-Kola Pop a sample of children drinks in a day. From the sample of children, the researcher finds that the five children consume the following numbers of the Super-Kola Pop, 1, 3, 3, 4, 4. To work out the standard deviation the researcher would need to do the following.

See the table below to see how each of the above steps are carried out.

- *Step 1*: Work out the mean of the numbers.

- *Step 2*: Subtract (take-away, minus) the mean average from each of the numbers, to gain deviations.

- *Step 3*: Square (times by itself) each of the deviations to get squared deviations.

- *Step 4*: Add all the squared deviations together to get the sum of the squared deviations.

- *Step 5*: Divide the sum of the squared deviations by the number of people in the sample minus 1 to find the variance.

- *Step 6*: Find the square root of the variance to compute the standard deviation.

Step 1: $1+3+3+4+4=15$, 15 divided by $5=3$, Mean $=3$			
Score	Mean	Deviation (*Step 2*)	Squared deviation (*Step 3*)
1	3	$1-3=-2$	$2 \times 2 = 4$
3	3	$3-3=0$	$0 \times 0 = 0$
3	3	$3-3=0$	$0 \times 0 = 0$
4	3	$4-3=1$	$1 \times 1 = 1$
4	3	$4-3=1$	$1 \times 1 = 1$
Step 4: $4+0+0+1+1=6$			
Step 5: 6 divided by $5-1$, 6 divided by $4=1.5$			
Step 6: Square root of $1.5 = 1.22$, *Standard deviation* $= 1.22$			

Exercise 4: Calculating the standard deviation

If you have a calculator, either a handheld one, or on the computer, work out the standard deviation for a second study the researcher carried out. The researcher wanted to know how many cans of Diet Super-Kola Pop a sample of children drink in a day. From the sample of children the research finds that five children consume the following numbers of cans, 1, 3, 4, 5, 7.

1. What is the standard deviation for this sample for drinking Diet Super-Kola Pop?

2. Which set of scores has the highest standard deviation, those for drinking Super-Kola Pop, or the set for the diet brand? What does the higher standard deviation for this set of scores mean?

Frequencies on SPSS for Windows

In this example, we are going to obtain the frequencies for one variable in the dataset: the variable Sex (number of males and females in the sample). Load up the dataset. We are going to produce a frequency table for the variable Sex. Click on **Analyse** pull-down menu, click on **Descriptive Statistics**, pull the mouse over to **Frequencies**, and click on this. Your screen will look like Figure 3.11.

Move the variable 'sex' from the left hand box into the **Variable[s]:** box by using the >. Then press **OK**. You should get a table like Table 3.3.

In this frequency table we are mainly interested in one, or perhaps two things. We are interested in the frequency of each level (Male and Female), and the percentage breakdown of each level. In this example we can see that the sample contains 29 males and 31 females. In terms of percentages (though this is seldom reported) 48.3 per cent of the sample were male, and 51.7 per cent of the sample were female.

Figure 3.11   Frequencies window.

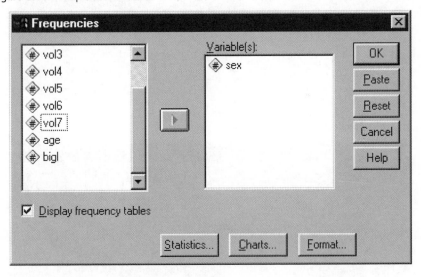

Table 3.3   Frequency table for Sex.

		Frequency	Per cent	Valid per cent	Cumulative per cent
Valid	Male	29	48.3	48.3	48.3
	Female	31	51.7	51.7	100.0
	Total	60	100.0	100.0	

Mean and standard deviations on SPSS for Windows

We are going to obtain the averages and standard deviation for one variable in the dataset: scores on a measure of Belief in Good Luck (Darke and Freedman, 1997). This is a continuous variable that measures the level to which an individual believes in good luck. Items include statements such as 'Luck plays an important part in everyone's life' (item a) and 'I consistently have good luck' (item g). Respondents are required to indicate the extent of their agreement, using a response format from (1) *Strongly disagree* to (6) *Strongly agree*.

To get the average statistics and standard deviation for this variable, click on the **Analyse** pull-down menu, click on **Descriptive Statistics**, then pull the mouse over to **Frequencies**, and click on this. The screen will look similar to the previous one for frequencies. Transfer the variable BIGL into the **Variable[s]:** box using the >. However, before pressing OK, click on the **Statistics...** box at the bottom of the window. The screen will look like Figure 3.12. Click in the boxes next to **Mean**, **Median** and **Mode** under Central Tendency, and click the box next to **Std. deviation** under Dispersion. Now press **Continue**, and then press **OK**. You will get a table that looks like Table 3.4.

Figure 3.12   Frequencies: Statistics.

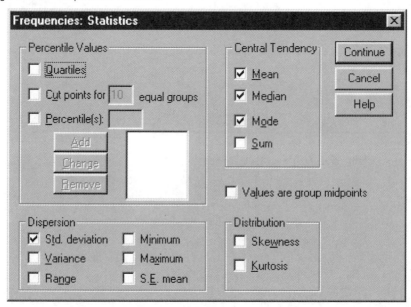

Table 3.4   Statistics for Belief in Good Luck.

N	Valid	60
	Missing	0
Mean		34.7333
Median		34.5000
Mode		31.00
Std. Deviation		8.2747

As you can see, the mean average is 34.73, the median is 34.50, the mode is 31.00 and the standard deviation is 8.2747.

Exercise 5: Obtaining frequencies, mean and standard deviation using SPSS for Windows

Load up the datafile. Using this data obtain a frequency table, mean average and standard deviation for the variable Age (age of respondents).

1. How old is the youngest person(s) in the sample?
2. How old is the oldest person(s) in the sample?
3. How many people are aged 31 in the sample?
4. What percentage of people are aged 29?
5. What is the mean, median, mode for age?
6. What is the standard deviation?

Graphs: who used all the pie charts?

On SPSS for Windows you can also provide graphical representations of variables. One advantage of descriptive statistics is that you can carry out a number of graphical representations. However, it is often a temptation to become over involved with graphs. One such graph is a pie chart. Pie charts on SPSS for Windows are brilliant, colourful and very impressive, particularly when a 3D selection is used. Yet, they are little used in research. Rather, we are going to concentrate on the two simple graphs (the bar chart and the histogram), not only because they are frequently used, but because one of them (histogram) will be used as an important building block in understanding further statistics.

Bar charts and histograms are ways of representing data visually. Bar charts are used for categorical data, and histograms are used for continuous data. We will, using the variables mentioned above (Sex and Belief in Good Luck), show how to obtain bar charts and histograms on SPSS.

Obtaining a bar chart on SPSS

Pull down **Graphs**, and click on **Bar...** and a new screen will appear. Click on **Simple**, and then **Define**. Move the variable Sex into the **Category Axis**, and press **OK**. You will now get a chart like Figure 3.13.

The variable levels (male and female) are plotted along the bottom (this is called the *x*-axis) and the frequency of each level is plotted up the side (this is called the *y*-axis). Notice how the bars are separate, indicating that the variable is categorical.

Obtaining a histogram on SPSS

Pull down **Graphs**, and click on **Histogram...** and a new screen will appear. Transfer the BIGL variable in the **Variable** box and press on **OK**. You will then get the output as in Figure 3.14.

The variable levels (possible scores) are plotted along the *x*-axis and the frequency of each level is plotted along the *y*-axis. Notice how the bars are together, indicating the variable is continuous.

Figure 3.13   A bar chart of the variable Sex.

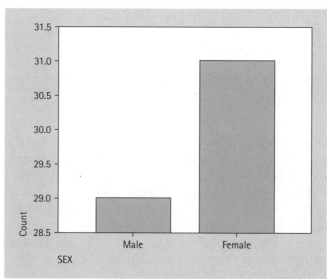

Figure 3.14   A histogram of the variable Belief in Good Luck.

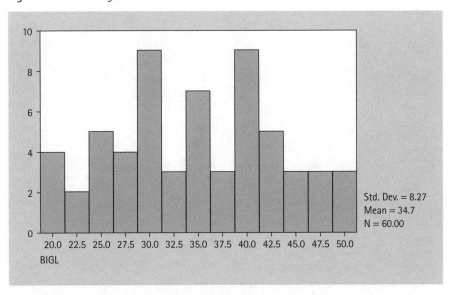

Exercise 6: Creating bar charts and histograms in SPSS for Windows

Plot a graphical representation of the Age variable and two other variables of your choice from the dataset. Be careful to choose the right sort of graph to suit the type of variable.

This is as far as we need to go at this time. In the next chapter we will start to explore the histogram more fully. This is because the histogram, and more importantly what it can demonstrate, holds the key to the inferential statistics we shall examine later in the book.

References

Darke, P.R. and Freedman, J.L. (1997). The Belief in Good Luck Scale. *Journal of Research in Personality*, **31**, 486–511.

Francis, L.J. and Stubbs, M.T. (1987). Measuring attitudes towards Christianity: from childhood to adulthood. *Personality and Individual Differences*, **8**, 741–743.

Hartley-Brewer, J. (1999). Fun-loving Britons splash out most on leisure. *The Guardian*, 25 November.

Lewis, C.A. (2000). The Depression–Happiness Scale. In J. Maltby, C.A. Lewis and A.P. Hill (Eds), *A handbook of psychological tests*, Vol. 2. Cardiff: Edwin Mellen Press.

McGreal, R. and Joseph, S. (1993). The Depression–Happiness Scale. *Psychological Reports*, **73**, 1279–1282.

Wann, M. (1999) Self-hatred and celery sticks. *The Guardian*, 29 March.

Answers to exercises

Exercise 1: Chapter energiser

Question 1. 1. 15   2. 13
Question 2. 1. 2, 3, 5, 7, 9, 17   2. 1, 2, 2, 7, 53, 62
Question 3. 1. 31, 17, 15, 12, 9, 5, 4   2. 53, 23, 11, 7, 2, 2
Question 4. 1. 1, 2, 3, **4**, 5, 8, 9; Middle number = 4
             2. 12, 13, 17, **18**, 19, 21, 23; Middle number = 18
Question 5. 1. 2
             2. 117
Question 6. 1. $1 + 7 + 6 + 3 + 3 + 4 = 24$; 24 divided by 6 = 4
             2. $23 + 27 + 18 + 12 + 20 = 100$; 100 divided by 5 = 20

Exercise 2: Reversing and computing scores on a scale

1. The items 'I think going to church is a waste of my time' and 'I think the Bible is out of date' will need to be reverse scored.

2. The respondent should score 16.

Exercise 3: Calculating the mean, median and mode

1. Mean = 4.8, median = 3, mode = 3.

2. The mean is higher, and is distorted due to the value 15.

Exercise 4: Calculating the standard deviation

1. Standard deviation = 2.23.

2. The standard deviation for the diet brand is higher, suggesting that the scores (number of cans) are more spread out around the mean.

Exercise 5: Obtaining frequencies, mean and standard deviation using SPSS for Windows

1. 19 years.

2. 52 years.

3. One.

4. 10 per cent.

5. Mean = 34.47, median = 34.00, mode = 33.00.

6. 6.72.

Exercise 6: Creating bar charts and histograms in SPSS for windows

1. You should have used a histogram to plot scores on the AGE variable.

CHAPTER 4

An introduction to inferential statistics

Although this may seem a paradox, all exact science is dominated by the idea of approximation.

(Bertrand Russell 1872–1970)

During the next two chapters, you will be introduced to a series of inferential statistics. This section is designed to give you a brief overview of the rationale behind the use of inferential statistical tests. You will be introduced to distributions, probability and error, and significance testing. At the end of the chapter, you should be able to make decisions about using parametric versus non-parametric tests, and the importance of probability values to determine the significance of results.

Exercise 1: Chapter energiser

Try the Arrow Numbers Puzzle in Figure 4.1. One concept we are going to introduce in this chapter is the idea that you may be able to draw conclusions from incomplete data. This puzzle is designed to put you in the right frame of mind for tackling these skills.

Distributions

In the last chapter you were shown how to create a histogram on SPSS for Windows (the distribution of scores for a continuous variable). Researchers have identified different types of distributions of a histogram to introduce an idea that is a cornerstone of statistics. This idea is based on different ways of describing distributions.

Figure 4.1   Arrow Numbers Puzzle.

Each number already in the grid shows the *sum* of the digits in the line whose direction is shown by the arrow.

Only one digit can be placed in each square. There are no zeros.

For each sum, each digit can only be used once – e.g. 8 *cannot be* completed with 4 + 4.

Can you complete the grid? You have been given one digit to start you off.

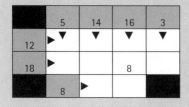

Figure 4.2   Negative and positive skewed distributions.

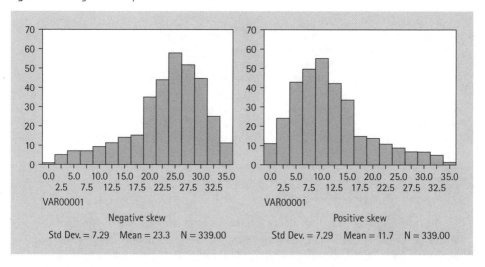

Histograms can be described as *skewed* (see Figure 4.2), either negatively (where scores are concentrated to the right) or positively (where scores are concentrated to the left). An example of a negatively skewed distribution may occur when high scores on the variable are highly desirable. Therefore, if researchers develop a 'kindness' scale (in which higher scores indicate a higher level of kindness) containing items such as 'I am kind to other people', and 'I am a very kind person', we might expect most respondents to view themselves as being kind, as opposed to being unkind. Therefore, a negative skew would emerge as people respond with high 'kindness' scores. An example of a positively skewed distribution of a variable is often found with measures of depression. Measures of depression tend to identify greater and greater degrees of severity of depression, so researchers often find that most respondents score low on depression, as most people are not regularly depressed, and fewer people have high scores (as the highest scores indicate severe depression and fewer respondents tend to be clinically depressed).

How do you know a distribution is skewed?

By obtaining a skew statistic using SPSS for Windows. Often it is difficult to tell from a distribution chart whether something is skewed. Therefore, you can use the skewness statistic in SPSS to gain an idea. You get the skewness statistic from the **Analyse** pull-down menu, click on **Descriptive Statistics**, and then on **Descriptives**. Once you have transferred the variable into the box, click on the **Options** button, and tick the box next to skewness. By pressing **Continue** and then **OK**, you will get the skewness statistic with your output. This statistic will be either positive or negative, indicating in which direction the skewness may be. There is a test to decide whether your skewness statistic means you have a skewed distribution or not, but this is vulnerable to the number of people you have in your sample. Most researchers use a general criterion of any statistics above 1.00 (either + or –) means your distribution is skewed. You will find this information useful when you are making decisions about parametric and non-parametric tests.

The final description is where the distribution of scores for a variable forms a 'normal distribution'. This is where scores are distributed symmetrically in a curve, with the majority of scores in the centre, then spreading out, showing increasingly lower frequency of scores for the higher and lower values (Figure 4.3).

Figure 4.3   Normal distribution.

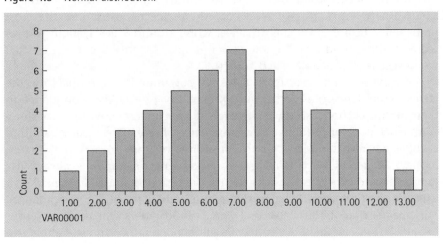

What is particularly notable about this is that researchers have found that many variables which measure human attitudes and behaviour follow a normal distribution curve. This finding is one of statistics' more interesting elements. Statisticians and researchers are often not certain why many variables fall into a normal distribution. They have just found that many attitudes and behaviours do so.

However, statisticians have noted that if scores on a variable show a normal distribution, this is potentially a powerful statistical tool. This is because we can then begin to be certain about how scores will be distributed in a variable (i.e. that many people's scores will be concentrated in the middle and few will be concentrated at either end). This certainty has given statisticians the impetus to develop ideas about statistical testing. These developing ideas centre on the concept of probability.

Probability

With the introduction of the National Lottery, the use of the word 'probability' has increased in society. The chances of winning the National Lottery Jackpot are thought to be about 14 million to 1, meaning that there is a very small probability that you will win. We can make a number of assertions about life based on probability. It is 100 per cent probable, if you are reading this sentence, that you have been born; there is a 50 per cent probability (1 in 2) that a tossed coin will turn up heads (50 per cent probability that it will turn up tails); and a 16.66 per cent probability (1 in 6) that a roll of a dice will end in a 6.

Some of the ideas about uses of probability in statistics have come from recognising that scores are often normally distributed. With normal distribution we are able to talk about how individual scores might be distributed. An example of this would be for a variable in which scores are normally distributed between 0 and 10. We would then expect most scores (the most probable) to be around 5, slightly fewer scores around 6 and 4, slightly fewer scores again around 7 and 3, and so on (8, 2; 9, 1), until the least expected scores (the least probable) are 0 and 10.

These expectations lead us to the key issue underlining probability, the idea of confidence. Researchers use probability to establish confidence in their findings. The reason why researchers are concerned with establishing confidence in their findings is a consequence of researchers using data that is collected from samples. Due to constraints of time to carry out studies, money, or accessibility to possible respondents, researchers always use sample data to generalise or infer (hence the generic name of *inferential* statistics for the tests we will use in the next few chapters) about a population. This means that there is always a chance researchers will make an error, because they never have access to the whole population, and therefore, can

Figure 4.4   Racing Form Card for the Fantasy Horse Novices' Hurdle.

3.40   2m7$^{1}/_{2}$
Fantasy Horse Novices' Hurdle (Class D) (4yo+)   2m7f110y
10 run Winner £2260   Going Good

No.	Horse	Wgt	Jockey
1	Liaw	10–12	C. Sabinti
2	Save Chip	10–12	S. Ockeram
3	Secret Button	10–12	R. Smith
4	Thursday Nights	10–12	U. Bar
5	Gareth's Cream Bun	10–14	L. Knight
6	Never find another horse better	10–12	J. Sullivan
7	Cheesy Snacks	10–12	C. Levi
8	Its just not Cricket	10–12	G. Roberts
9	Dirty Leads	10–12	C. Connille
10	Its Upside Down	10–12	W. Cooper
11	Tough Edward	10–7	U.R.P Able

Betting Forecast: 100/30 Secret Button, 2/1 Save Chip, 7/2 Liaw, 6/1 Gareth's
Cream Bun, 10/1 Its just not Cricket, 14/1 Dirty Leads, 16/1 Tough Edward,
18/1 Cheesy Snacks, 16/1 Thursday Nights, 25/1 Its Upside Down, 33/1 Never
find another horse better.

never be certain of how every possible respondent would have scored on a
variable. However, because researchers find that variables often form a normal distribution, they can use samples to generalise about populations with
confidence. Researchers try to establish confidence by talking about their
findings with regard to probability. An example of how they do this can be
seen in the horseracing form card shown in Figure 4.4.

At the bottom of this form is the betting forecast. As you can see *Secret
Button* is the favourite at 100/30, with *Never find another horse better*, being
least favourite, at 33/1. In a similar way that bookmakers are suggesting that
it is probable that *Secret Button* will win, and *Never find another horse better*
will probably not win, researchers use probability to grade findings as more
or less probable. However, unlike bookmakers, researchers use a criterion
to decide whether something is probable or not. The way this is done is
through significance testing.

Significance testing

Significance testing is a criterion, based on probability, that researchers use
to decide whether two variables are related. Remember, as researchers always
use samples, and because of the possible error, they use significance testing
to decide whether the relationships observed are real, or not.

Researchers are then able to use a criteria level (significance testing) to decide where their findings are probable (confident of their findings) and not probable (not confident of their findings). This criterion is expressed in terms of percentages, and their relationship to probability values. If we accept that we can never be 100 per cent sure of our findings, we have to set a criterion of how certain we want to be of our findings. Traditionally, two criteria are used. The first is that we are 95 per cent confident of our findings, the second is that we are 99 per cent confident of our findings. This is often expressed in another way. Rather, there is only 5 per cent (95 per cent confidence) or 1 per cent (99 per cent confidence) probability that we have made an error. In terms of significance testing these two criteria are often termed the 0.05 (5 per cent) and 0.01 (1 per cent) significance levels.

In the next few chapters you will be using tests to determine whether there is a *significant* association/relationship between two variables. These tests always provide a probability statistic, in the form of a value; e.g. 0.75, 0.40, 0.15, 0.04, 0.03, 0.002. Here, the notion of significance testing is essential. This probability statistic is compared against the criteria of 0.05 and 0.01 to decide whether our findings are significant. If the probability value (p) is less than 0.05 ($p < 0.05$) or less than 0.01 ($p < 0.01$) then we conclude that the finding is significant.[1] If the probability value is more than 0.05 ($p > 0.05$) then we decide that our finding is not significant. Therefore, we can use this information in relation to our research idea, and we can determine whether our variables are significantly related, or not. Therefore, for the probability value stated above:

- The probability values of 0.75, 0.40 and 0.15 are greater than 0.05 (> 0.05) and these probability values are not significant at the 0.05 level ($p > 0.05$).

- The probability values of 0.04 and 0.03 are less than 0.05 (< 0.05) but not less than 0.01, so these probability values are significant at the 0.05 level ($p < 0.05$).

- The probability value of 0.002 is less than 0.01 (< 0.01), therefore this probability value is significant at the 0.01 level ($p < 0.01$).

An analogy of significance testing to research is the use of expert witnesses with evidence in court cases. In a court case, an expert witness is used to comment on a piece of evidence to help the jury draw a conclusion about the accused. In the same way, the researcher uses significance testing (the expert witness) to help to determine whether the finding (evidence) is significant, or not (the jury conclusion).

---

[1] If the probability statistic is below 0.01 (e.g. 0.002) then we don't concern ourselves with mentioning it is below 0.05 because we are more confident of our finding (99 per cent).

Hypothesis testing

So far in this book we have looked at research as a way of answering research questions. However, one thing you have to be aware of is that sometimes a research question is referred to as a hypothesis. Using hypotheses is the same as using research questions. However, there is some formal terminology associated with using hypotheses to describe the type of research question proposed.

In using hypotheses, you are asked to state *formally* the outcomes expected in the research, in terms of the relationship between the variables measured, and whether you expect the relationship to be significant. The first formal statement of outcomes relates to making a *null hypothesis*. With the null hypothesis you are suggesting that there will be *no* relationship between the variables you are measuring. Therefore, if a researcher is looking at the relationship between happiness and depression, the null hypothesis would be that there would be no significant relationship between happiness and depression.

The second statement is concerned with an *alternative hypothesis*. With the alternative hypothesis you are suggesting that there *will* be a relationship among the variables you are measuring. Therefore, if a researcher is looking at the relationship between happiness and depression, the alternative hypothesis would be that there would be a significant relationship between happiness and depression.

When you are using, or are required to use, hypotheses to state your research questions, you must always state both the alternative and null hypotheses.

Type I and Type II errors

Within the research literature you may often come across references to Type I and Type II errors. Type I and Type II errors refer to the way that researchers can classify the possible errors that can be made when using significance testing. In this chapter we have suggested that all findings, due to using samples, are open to error, and significance testing is used to establish confidence in our results (remember, researchers can never be 100 per cent sure of any findings, rather, we refer to 95 per cent [$p = 0.05$] and 99 per cent [$p = 0.01$]) confidence). Simply put, Type I error refers to an occasion when a researcher may conclude that the findings are significant; when in actuality the findings are not significant. A Type II error refers to an occasion when a researcher may conclude that a finding is not significant; when in reality the findings are significant.

Exercise 2: Using probability

Using the following probability values, decide whether the statistic is significant or not significant. Then decide, if the result is significant, which level of significance the statistic is at (0.05 or 0.01).

1.   0.060
2.   0.030
3.   0.500
4.   0.002
5.   0.978

Significance testing with statistical tests

In the next two chapters you will be introduced to a number of statistical tests. In Chapter 5 you will be introduced to a series of *parametric* tests. In Chapter 6 you will be introduced to a series of *non-parametric* tests.

Consider the decision-making table for choosing statistical tests that you studied in Chapter 1 (Figure 4.5). As you can see, the second decision-making question is whether your continuous data should be used in a parametric test, or a non-parametric test. This decision is made when at least one of your variables can be described as continuous data.

Criteria for using parametric tests

Many people have different views about the criteria for deciding whether continuous data can be used in a parametric test. The traditional view is that:

- *Rule 1*: The scores on any continuous data must be interval or ratio data (or continuous if you are using the categorical/discrete/continuous distinctions between variables).
- *Rule 2*: The scores on any continuous data must be normally distributed.
- *Rule 3*: If two sets of continuous scores are being used or compared, they must have similar variances, standard deviations (known as homogeneity of variance).

Therefore, if your data fulfils all criteria, you use a parametric statistical test.

Criteria for using non-parametric tests

The criteria for using a non-parametric test reflect these three rules. Therefore, if your continuous data does *not* fulfil this criterion you *do not* use a parametric statistical test; instead you use a non-parametric test. So for *Rule 1*,

Figure 4.5   Decision-making table for choosing statistical tests.

Question 1 What combination of variables have you?	Which test to use	Question 2 Should your continuous data be used with parametric tests or non-parametric tests?	Which test to use	Question 3 How many levels has your categorical data?	Which test to use
Two categorical	Chi-square				
Two separate continuous	Go to Question 2	Parametric	Pearson		
		Non-parametric	Spearman		
Two continuous which is the same measure administered twice	Go to Question 2	Parametric	Related *t*-test		
		Non-parametric	Wilcoxon sign-ranks		
Two continuous which is the same measure administered on three occasions or more	Go to Question 2	Parametric	ANOVA (within subjects)		
		Non-parametric	Friedman test		
One categorical and one continuous	Go to Question 2	Parametric	Go to Question 3	2	Independent-samples *t*-test
				3 or more	ANOVA (between subjects)
		Non-parametric	Go to Question 3	2	Mann–Whitney U
				3 or more	Kruskal–Wallis

if your data is not interval or ratio data then you should be using a non-parametric test. For example, consider an ordinal variable which asks the respondent how their anxiety might be scored: 1 = Not at all, 2 = Sometimes, 3 = Often, 4 = Always. It is sometimes argued that this scoring does not reflect numerical properties; as the score of 4, given for responding 'Always' to the question, does not represent a behaviour that is twice as much as the score of 2, given for 'Sometimes' (Always is not twice as much as Sometimes). Similar principles apply for *Rules 2 and 3*. If your continuous data is *not* normally distributed (Rule 2) or if two sets of continuous scores are being used or compared and they *do not* have similar standard deviations (Rule 3) then a non-parametric test should be used.

Some practices relating to criteria for choosing parametric and non-parametric tests

It is worth noting that some people don't always use these rules, or will vary in the application of these rules. Even when the data doesn't comply with the rules, researchers may use parametric tests for a number of good reasons, including:

- The statistics test has been previously used by other researchers to look at the variable. This is sometimes done to provide consistency in the research literature.

- Sometimes researchers will assume their continuous data is normally distributed because they have collected data with adequate sampling techniques, and therefore have data that is in general representative of a normally distributed population (applicable to Rule 2).

- The scale used to measure the continuous variable is a well-established reliable and valid measure of that variable which has been shown by previous research (among larger samples) to demonstrate a normal distribution of scores (applicable to Rule 2).

- Although some continuous variables may not comprise real numbers (e.g. ordinal data), we can make the assumption that they *are* real numbers because researchers are assigning these values to responses (applicable to Rule 1).

- Because their lecturer/teacher has told them so. Lecturers and teachers are wise people and should be fully listened to on all occasions (mainly because they are tending to mark your work). Therefore, lecturers/teachers may suggest you use a certain version of the test because of wider learning reasons, or for reasons of being consistent with the literature.

What you need to be aware of most is that researchers do vary in practice. This is often a source of confusion for students of statistics, and the best strategy you can employ is just to be aware that people do vary in practice. Usually, there is little point in adopting one position only, because often you will work with colleagues, complete reports for employers/teachers/lecturers, or submit papers to academic journals, who will insist you adopt a different tactic. Ideally, you should be aware that different valid practices exist, and be able to employ and engage with these different practices when needed.

It is best to use these various rules and practices as guidelines to decide whether you use parametric or non-parametric versions of tests. However, overall you should remember that this shouldn't be viewed as a big problem. The purpose is not that you should develop a concept of parametric versus non-parametric tests. Rather, statisticians have provided us with different ways of solving problems. Therefore, there is nothing to worry about if you find that the continuous variables you have measured do not show parametric properties; you simply use non-parametric tests as an alternative.

A summary of the strategy for choosing either a parametric or a non-parametric test

The approach mentioned above is a useful tool in deciding whether to use parametric or non-parametric tests. Simply remember that non-parametric tests are used when you can show the continuous variables you are using are not displaying the properties associated with a normally distributed numerical variable, or, in the case of two sets of continuous variables being compared, homogeneity of variance. Therefore, if the continuous variable you are using isn't normally distributed, but shows a skewed distribution, or if your variable isn't interval data, but you want to use the variable as a continuous variable, then you simply use a non-parametric version of the test.

Therefore, in using the decision-making table for choosing statistical tests you need to decide whether your continuous data can be used in parametric tests or non-parametric tests. First assume that you are going to use the parametric tests. However, then consider a number of factors. If the data is skewed (the most common reason), does not comprise real numbers (or you don't want to assume they have the properties of real numbers), or if you are using two continuous variables and they have unequal standard deviations (don't show homogeneity of variance), treat the continuous data as being suitable for non-parametric versions of tests. Further, if there is a reason for treating the data as parametric, because you are using an established scale, or to be comparable with previous research, then state the reason why and use the parametric test.

Question 1 revisited in the decision-making chart

In the last section it is worth noting that some distinctions in the criteria for whether to use parametric or non-parametric tests centred on distinctions between ordinal/discrete data and interval/ratio/continuous data. At this stage, it is worth returning to the first question in the decision-making chart, to help understand further some of the differing practices among researchers. This question asks you to decide the combination of variables you have (e.g. two categorical, two continuous, one categorical or one continuous). In Chapter 2 we noted that researchers differ in the way that they classify ordinal (ranked ordered categories) and discrete data (discrete data does not allow for decimal points). Some would describe this data as categorical and some would describe this data as continuous; this is therefore a possible source of confusion for you. However, in summary we see three different systems that researchers may use with data described as ordinal or discrete.

1. Treat all ordinal and discrete data as categorical tests (because the data is not considered as comprising real numbers). Though it is worth noting that this is usually done when there are only a few levels of the variable – e.g. Good (1), Fair (2), OK (3), Not so good (4) – and not when there are many levels (around six or more).

2. Treat all ordinal and discrete data as continuous, but always use *non-parametric* statistical tests (because the data is not considered as comprising real numbers, and therefore, cannot show normal distribution properties).

3. Treat all ordinal and discrete data as continuous, but always use *parametric* statistical tests (because an assumption is made that we are assigning numerical values to responses).

4. Treat all ordinal and discrete data as continuous, but check whether the data is suitable for using with parametric or non-parametric tests.

We would suggest that you use the fourth system: treat ordinal/discrete data as continuous, and check whether your data fulfils the other criteria for parametric testing. For example, examine whether your continuous data demonstrates a normal distribution, and if two sets of continuous scores are being used or compared, whether they have similar standard deviations. If they do not show these properties, then you decide that the data is suitable for non-parametric testing.

It is worth noting that researchers will still use the other systems, but again, just be aware that different valid practices exist, and be able to employ and engage with these different practices.

Revisiting the decision-making chart

We have described the two processes relating to Questions 1 and 2 of the decision-making chart and we need to summarise the main points. Though there is differing practice we suggest you use the following rules:

- When answering Question 1 make the distinction between whether each variable is categorical or continuous (treat nominal/categorical data as categorical; treat interval/ratio/continuous/ordinal/discrete as continuous).

- When answering Question 2 employ the rule around criteria for parametric testing, and use descriptive statistics (e.g. histograms/skewness statistics) to decide whether your data is suitable for parametric or non-parametric testing.

Bringing significance testing and the decision-making table together

In summary, we have covered significance testing and revisited the decision-making table with types of variable and parametric and non-parametric testing. We will combine these two elements together to show how you can form a basis for significance testing using the decision-making chart (Figure 4.5).

Exercise 3: Using significance testing and the decision-making table together

Using the decision-making table for choosing statistical tests, answer the following questions. For each question decide: (a) what statistical test the researcher should use; and (b) whether the researcher's results are significant or not.

1. Researcher A has a measure of sex (males and females) and exam scores on a maths test to see if men and women differ in their ability in maths. The researcher has decided that sex is categorical and exam scores on a maths test is a continuous variable. Using descriptive statistics, the researcher has decided that the continuous variable demonstrates a normal distribution.
   (a) What statistical test should the researcher use?
   (b) The researcher finds that after running the test, the probability value is 0.63. Are the findings that are going to be reported by the researcher significant, or not significant?

2. Researcher B has administered measures of anxiety before and after an exam to see if the anxiety levels change before and after an exam. The researcher has decided that these are both continuous variables, but has found on this occasion, that both variables show a negatively skewed distribution.
   (a) What statistical test should the researcher use?
   (b) The researcher finds that after running the test, the probability value is 0.04. Are the findings that are going to be reported by the researcher significant or not significant?

Answers to exercises

Exercise 1: Chapter energiser

See Figure 4.6 for the solution to the Arrow Numbers Puzzle.

Figure 4.6 Solution to the Arrow Numbers Puzzle.

	5	14	16	3
12	▶ ▼ 1	▼ 3	▼ 6	▼ 2
18	▶ 4	5	8	1
	8	▶ 6	2	

Exercise 2: Using probability

1. This is not significant.
2. This is significant at the 0.05 level.
3. This is not significant.
4. This is significant at the 0.01 level.
5. This is not significant.

Exercise 3: Using significance testing and the decision-making table together

1. (a) An independent-samples *t*-test should be used.
   (b) The findings are not significant ($p > 0.05$).
2. (a) A Wilcoxon sign-ranks test should be used.
   (b) The findings are significant ($p < 0.05$).

An introduction to parametric tests:
the relationship between belief in good luck, happiness and neuroticism

The simplest schoolboy is now familiar with facts for which Archimedes would have sacrificed his life.

(*Ernest Renan*, Souvenirs d'enfance et de jeunesse)

Within this session you will learn the rationale, the procedure and the interpretation for five parametric tests:

- Pearson product–moment correlation coefficient (used when you have two different continuous variables that you have decided can be used in a parametric test).
- Related *t*-test (used when you have the same continuous variable, administered on two occasions, and you have decided the data is suitable for use in a parametric test).
- Independent-samples *t*-test (used when you have one categorical variable with two levels, and one continuous variable that you have decided can be used in a parametric test).
- Analysis of variance – between subjects (used when you have one categorical variable with three levels or more, and one continuous variable that you have decided can be used in a parametric test).
- Analysis of variance – within subjects (when you have the same continuous variable administered on three occasions, that you have decided can be used in a parametric test).

In this chapter, aspects of SPSS for Windows will be illustrated using the dataset 'Datafile.sav', so ensure you load up this dataset when using SPSS for Windows.

Exercise 1: Chapter energiser

Try the puzzle in Figure 5.1. The first statistical test introduced in this chapter is a correlation statistic. This puzzle is designed to put you in the frame of mind for tackling the concepts underlying this statistical test.

Figure 5.1   Mathematical Puzzle.

How to solve: *Record in this grid all the information obtained from the clues, by using an 'X' to show a definite 'no' and a '/' to show a definite 'yes'. You'll find that this narrows down the possibilities, and may reveal some new definite information. So now you re-read the clues with these new facts in mind to discover further positive/negative relationships. Transfer these to all sections of the grid, thus eliminating all but one possibility, which must be the correct one.*

The FA Cup has narrowed down its teams to the last eight. From the clues given below, can you work out which teams were drawn to play against each other (showing the first and second team to be drawn), on what dates the matches took place, and what was the score.

1. Blackpool isn't the team who played Chelsea, and Chelsea beat their opponents 2–0.

2. Newcastle United played Sheffield Wednesday but not on 26 November.

3. Liverpool, who drew 1–1, played immediately before Arsenal, who didn't play in December.

4. The match on 19 November showed a score of 1–0.

5. Manchester United, who played on 5 November, didn't play Chelsea or Grimsby, and didn't lose 1–3.

	Rotherham	Grimsby	Arsenal	Chelsea	Sheff Wed	5 Nov	12 Nov	19 Nov	26 Nov	3 Dec	2–1	1–1	1–0	2–0	1–3
Manchester Utd															
Liverpool															
Blackpool															
Sunderland															
Newcastle Utd															
2–1															
1–1															
1–0															
2–0															
1–3															
5 Nov															
12 Nov															
19 Nov															
26 Nov															
3 Dec															

First team drawn	Second team drawn	Date played	Score

Hint: Once you have a tick in a row/column, you can eliminate all other squares in that row/column

Parametric tests

Parametric tests are used when the continuous data meets the assumptions needed for using continuous data in a parametric test.

Parametric test 1: Pearson product–moment correlation coefficient

The aim of the correlation coefficient is to determine the relationship between two separate continuous variables.

Imagine two variables, *amount of chocolate eaten* and *weight*. It is thought that chocolate contains a lot of fat, and that eating a lot of fat will increase your weight. Therefore, it could be expected that people who eat a lot of chocolate would also weigh a lot. If this were true then the amount of chocolate you eat and your weight would be positively correlated. The more chocolate you eat, the more you should weigh.

Conversely, a negative correlation would represent a process by which scores on one variable rise, while scores on the other variable decrease. An example of this would be the amount of exercise taken and weight. It is thought that taking exercise will usually lead to a decrease in weight. If this were true then the amount of exercise you take, and your weight would be negatively correlated. The more exercise you take the less you might weigh.

Finally, some variables might not be expected to show a correlation with each other. For example, the number of hot meals you have eaten and the number of times you have visited the zoo. Usually, we expect that there would be no logical reason why eating hot meals and zoo visiting would be related, so that eating more hot meals would mean you would visit the zoo more, or less (or vice versa). Therefore, you would expect the number of hot meals you have eaten and the number of times you have visited the zoo *not* to show any correlation. We use parametric tests and significance testing to determine whether a correlation occurs between two variables.

An easy way of illustrating this is through scattergrams. Scattergrams are graphs that plot individuals' scores for one variable against individuals' scores on another variable. Consider the three graphs showing scattergrams of scores against two variables, A and B, shown in Figure 5.2.

Figure 5.2(a) shows how scores will be when you have a 'positive relationship'. The low values on one variable tend to correspond with low values on the other variable, and high values on one variable correspond with high values on the other. A positive relationship is shown by points plotted moving from the lower left hand corner up and across to the upper right hand corner of the chart. You can also have 'negative relationships' between

Figure 5.2 Examples of positive, negative and no relationship between two variables.

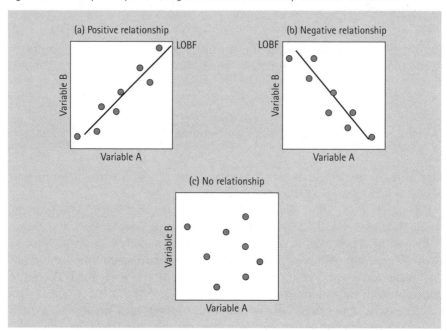

two variables where the low values on one variable tend to go with high values on the other variable and vice versa. A scattergram of a negative relationship would be similar to Figure 5.2(b), with the plotted points moving from the upper left hand corner down and across to the lower right hand corner of the chart. Finally, Figure 5.2(c) depicts what scores might be with no relationship: a more or less random scatter of plots with the plots showing no clear direction, going neither up nor down.

A correlation coefficient can be pictured as the single straight line ('line of best fit'; LOBF on the examples) which will come closest to all of the points plotted on a scattergram of association between two continuous variables. However, using the correlation statistic is more informative than this. The correlation coefficient provides a statistic that tells you the direction, strength, and significance of the relationship between two variables.

As a test, the correlation coefficient can take values ranging from +1.00 through 0.00 to −1.00.

- A correlation of +1.00 would be a 'perfect' positive relationship.

- A correlation of 0.00 would be no relationship (no single straight line can sum up the almost random distribution of points).

- A correlation of −1.00 would be a 'perfect' negative relationship.

Pearson's product–moment correlation coefficient in SPSS for Windows

In the following example we will show how a correlation works by using an aspect of research that looks at the relationship between belief in good luck and psychological well-being. Read the article given in Figure 5.3.

Figure 5.3   'Belief in Good Luck'.

Psychological Reports, 1999, 85, 971–972. © Psychological Reports 1999

Relationship between Belief in Good Luck and General Health

LIZA DAY, JOHN MALTBY, AND ANN MACASKILL
School of Health and Community Studies
Sheffield Hallam University, England

Summary: 62 undergraduate university students were administered the 12-item Belief in Good Luck Scale of Darke and Freedman and the General Health Questionnaire of Goldberg and Williams. Scores on belief in good luck showed a significant correlation of –0.29 with anxiety and –0.35 with depression but correlations were not significant for somatic symptoms (0.15) and social dysfunction (0.15).

Within the literature, there are traditionally two psychological explanations of luck. The first is luck as an external, unstable factor within social events or achievement outcomes (Rotter, 1955, 1966; Weiner, Frieze, Kulda, Reed, Rest, & Rosenbaum, 1972). Here, luck is perceived as uncontrollable and having little influence on future expectations as well as a rational belief. Thus, luck is thought to have no influence on the psychological well-being and health of the individual. The second explanation is luck as a personal attribute, as an internal and stable factor (Darke & Freedman, 1997b). In this explanation luck is often seen as an irrational belief, typically considered maladaptive (Ellis, 1971, 1973) and is thought to have a detrimental effect on individual's psychological well-being and health (Rotter, 1966; Seligman, 1975).

Some researchers, however, have begun to re-evaluate the latter assumption that beliefs surrounding luck are necessarily maladaptive and instead have speculated that they may be adaptive as these positive illusions can lead to feelings of confidence, control, and optimism (Taylor & Brown, 1988; Darke & Freedman, 1997a). As part of this revaluation, Darke and Freedman (1997a) developed a 12-item Belief in Good Luck Scale to measure such beliefs and assess their implications for perceptions of control. They reported that items on the scale showed a good internal reliability ($\alpha = 0.85$) and significantly correlated with locus of control scores but not with scores on self-esteem, desire for control, and achievement motivation. These findings provided both convergent and discriminant validity for the scale. Despite the development of the scale, there has been no examination of whether belief in good luck is adaptive in psychological well-being and health so an examination between the relationship in belief in good luck, psychological well-being and health is required.

Undergraduate students (14 men, 38 women, and 10 undisclosed) of ages 18 to 44 years (*Mean = 21.4, SD = 5.24*) were administered the 12-item Belief in Good

continued

Figure 5.3 (*continued*)

Luck scale (Darke & Freedman, 1997a) and the General Health Questionnaire (Goldberg & Williams, 1991). Items such as 'I consider myself to be a lucky person' were scored on a 6-point scale anchored by 'strongly disagree' and 'strongly agree'. The General Health Questionnaire contains four subscales that measure aspects of general health, Severe Depression, Anxiety, Somatic Symptoms, and Social Dysfunction.

A significant negative Pearson product–moment correlation coefficient obtained for higher scores on the Belief in Good Luck scale and lower scores on the Severe Depression ($r = -0.35$, $p < 0.01$) and Anxiety ($r = -0.29$, $p < 0.05$) measures of the General Health Questionnaire. No significant relationship was found between scores on Belief in Good Luck and Somatic Symptoms ($r = -0.04$, $p > 0.05$) and Social Dysfunction ($r = 0.15$, $p > 0.05$).

This suggests that belief in good luck is associated with less depression and anxiety. Further, belief in good luck may indicate better psychological well-being and is not maladaptive as is traditionally accepted for irrational beliefs. Together with the research by Darke and Freedman (1997a), we suggest that believing in good luck may be an adaptive behaviour. For instance, belief in good luck may provide an important means of coping with the unforeseen events that happen by allowing individuals to remain optimistic when it is impossible to exercise direct control over the circumstances. It must be remembered, however, that the General Health Questionnaire is a self-report measure and, although the findings show that subjects who believe in good luck are more optimistic, there is also a possibility that they may tend to exaggerate how little depressed and anxious they are. Researchers may examine the relationship between these concepts. Notwithstanding these speculations, the findings suggest reconsidering whether beliefs in luck are necessarily maladaptive.

REFERENCES

Darke, P.R., & Freedman, J.L. (1997a) The Belief in Good Luck Scale. *Journal of Research in Personality*, 31, 486–511.

Darke, P.R., & Freedman, J.L. (1997b) Lucky events and belief in luck: paradoxical effects on confidence and risk-taking, *Personality and Social Psychology Bulletin*, 23, 378–388.

Ellis, A. (1971) *Reason and emotion in psychotherapy*. New York: Lyle Stuart.

Ellis, A. (1973) *Humanistic psychotherapy. The rational-emotive approach.* New York: McGraw-Hill.

Goldberg, D., & Williams, I. (1991) *A user's guide to the General Health Questionnaire*. London: NFER-Nelson.

Rotter, J.B. (1955) The role of the psychological situation in determining the direction of human behaviour. In M.R. Tones (Ed.), *The Nebraska symposium on motivation*. Vol. 3. Lincoln, NE: University of Nebraska. Pp. 245–249.

Rotter, J.B. (1966) Generalized expectancies for internal versus external control of reinforcement. *Psychological Monographs*, 80, No. 1 (Whole No. 609).

Seligman, M.E.P. (1975) *Helplessness: on depression, development, and death.* San Francisco, CA: Freeman.

Taylor, S.E., & Brown, D. (1988) Illusions and well-being: a social psychological perspective on mental health. *Psychological Bulletin*, 116, 21–27.

Weiner, B., Frieze, I., Kukla, A., Reed, L., Rest, S., & Rosenbaum, R.M. (1972) Perceiving the causes of success and failure. In E.E. Tones, D.E. Kanouse, H.H. Kelley, R.E. Nisbett, S. Valins, & B. Weiner (Eds), *Attribution: perceiving the causes of behavior.* New York: General Learning Press. Pp. 95–120.

Accepted November 8, 1999.

(Reproduced with permission of authors and publisher from: Day, L., Maltby, J. & Macaskill, A. Relationship between belief in good luck and general health. *Psychological Reports*, 1999, 85, 971–972.
© Psychological Reports 1999)

In summary, Day *et al.* (1999) found that belief in good luck was significantly negatively associated with depression and anxiety, but not with somatic symptoms and social dysfunction. This means, the higher people score on belief in good luck the lower they tend to score on depressive symptoms and anxiety. People who believe in good luck tend to be less depressed and anxious. Day *et al.* (1999) suggest further investigation of this.

One way of providing further support for this idea, is to examine the relationship between belief in good luck and other indicators of depression. One example is the depression–happiness measure mentioned in Chapter 4. The Depression–Happiness Scale is a self-report questionnaire containing 25 items measuring aspects of happiness and unhappiness (depression). Of these, 12 of the items concern positive thoughts and feelings, and 13 of the items concern negative thoughts and feelings. Respondents are asked about how they have felt in the past seven days, and asked to rate each item on a four-point scale: *Never* (0), *Rarely* (1), *Sometimes* (2), and *Often* (3). The 13 negative items are reverse scored so that possible scores on the scale can range between 0 and 75. Higher scores on the scale indicate a higher frequency of positive thoughts, feelings, and bodily experiences, i.e. happiness (Lewis, 2000).

Therefore, as depression and happiness are opposite constructs, and Day *et al.* (1999) suggest that there is a negative correlation between Belief in Good Luck and Depression, we might expect to find a significant positive correlation between Belief in Good Luck and Depression.

Let us test this with the dataset (datafile.sav). Load up the dataset and pull down the **Analyse** menu, choose **Correlate** and then **Bivariate**. The window as shown in Figure 5.4 will come up. Then select the two variables you wish

Figure 5.4   Bivariate Correlations window.

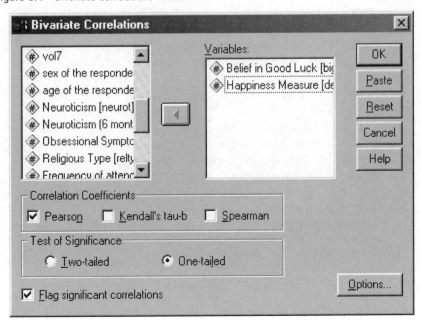

to correlate by transferring their names into the **Variables** box (Belief in Good Luck = BIGL; Happiness measure = DEPPHAPP).

Click on **Pears<u>on</u>** in the **Correlation Coefficients** box, and click the **One-tai<u>led</u>** circle in the **Test of Significance** box. Then run the correlation by clicking on **OK**. You will get an output which looks like Table 5.1.

Table 5.1 Correlation output.

		Belief in Good Luck	Happiness Measure
Belief in Good Luck	Pearson Correlation	1.000	0.403**
	Sig. (1-tailed)	.	0.001
	N	60	60
Happiness Measure	Pearson Correlation	0.403**	1.000
	Sig. (1-tailed)	0.001	.
	N	60	60

** Correlation is significant at the 0.01 level (1-tailed).

One- and two-tailed hypothesis testing

While null and alternative hypotheses involve making statements about whether there is a significant relationship (alternative hypothesis) or non-significant relationship (null hypothesis) between two variables, one-tailed and two-tailed tests involve making statements regarding the expected direction of the relationship between the two variables you are measuring. With a one-tailed hypothesis the researcher would make a statement regarding the specific direction of the relationship. With a two-tailed hypothesis no statement is made regarding the expected relationship. We can illustrate this with the present example, because we believe there is going to be a significant *positive* (this is the expected direction of the relationship) relationship between Belief in Good Luck and Happiness. A two-tailed hypothesis may be that we might expect a significant relationship, but we are unsure of the final direction of the relationship (here, the researcher would state that there is an expected significant relationship between the two variables). As we can see from the present example, by making a specific prediction about the direction of the relationship between Belief in Good Luck and Happiness, these ideas are incorporated into significant testing.

We are going to visit some further ideas of one-tailed and two-tailed hypothesis testing in the next section.

This output (Table 5.1) contains all the information we need for an interpretation of whether there is a relationship between Belief in Good Luck and Happiness. The important rule to remember when interpreting and writing test results is to *Describe* and then *Decide*. That is, describe what is happening within the findings, and then decide whether the result is significant.

Using Describe and Decide to interpret the test

From the output, you will need to consider three things:

- *Pearson correlation*. The statistical test statistic: it is important to note whether the statistic is positive or negative. This indicates whether the relationship between the two variables is positive (positive number, though SPSS doesn't print a +) or negative (represented by a – sign).
- The *Sig. (2-tailed)*. The significance level. This is the probability level given to the current findings.
- The significance level, found in the *Sig. (2-tailed)* bullet point above, and whether this figure suggests that the relationship between the two variables is significant or not. Remember, if this figure is below the p = 0.05 or p = 0.01 criteria, then the finding is significant. If this figure is above 0.05, then the finding is not significant.

The correlation between the variables Belief in Good Luck (BIGL) and Happiness (DEPPHAPP) is 0.403. This tells us that there is a positive relationship between Belief in Good Luck and Happiness (if it were negative the statistic would have a minus sign [–] in front of it). The significance level is p = 0.001. This is below 0.01. Therefore, we can conclude that there is a significant positive relationship between Belief in Good Luck and Happiness. This means that the more people believe in good luck, the happier they tend to be. We have therefore found further support for the findings of Day *et al.* (1999), that Belief in Good Luck is associated with better psychological well-being.

Using Decide and Describe to report the Pearson product–moment correlation coefficient

The next stage is that you will need to report these statistics. There is a formal way of reporting the Pearson product–moment correlation coefficient. There are two elements in this. The first element is a formal statement of your statistics must include:

- The test statistic. Each test has a symbol for its statistic. The Pearson product–moment correlation has the symbol *r*. Therefore, in writing your results you must include what *r* equals. In the example *r* = 0.403.

- The degrees of freedom. This is a figure that is traditionally reported (though it is worth noting that it is not always reported). For the Pearson product–moment correlation coefficient the degrees of freedom equal the size of your sample minus 2. The minus 2 represents minus 1 for each set of scores, the set of scores for Belief in Good Luck and the set of scores for Happiness. This is placed between the r and the = signs and placed in brackets. Here the degrees of freedom are 58 (size of sample = 60, minus 2 = 58). Therefore $r(58) = 0.403$.

- You must report the probability. This relates to whether your probability value was below 0.05 or 0.01 (significant) or above 0.05 (not significant). Here, you use less than (<) or greater than (>) the criteria level. You state this criterion by stating whether $p < 0.05$ (significant), $p < 0.01$ (significant) or $p > 0.05$ (not significant). In the example above, as $p = 0.001$, we would write $p < 0.01$ and place this after the reporting of the r value. Therefore, with our findings, $r(58) = 0.403$, $p < 0.05$.

What are degrees of freedom?

You will use degrees of freedom a lot with the hand-worked examples, but not if using SPSS for Windows. This is because 'degrees of freedom' is a concept that you use to help you decide whether your result is significant with hand-worked examples, but SPSS calculates this automatically.

Many authors rightly shy away from explaining what degrees of freedom are. However, one of the best explanations of how to conceptualise degrees of freedom is presented by Clegg (1983). Here, Clegg suggests you imagine two pots, one of coffee and one of tea, but neither are labelled. Clegg asks the question, 'How many tastes do you require before you know what is in each pot?'

The answer is one, because once you have tasted one, you not only know what is in that one, but also what is in the other. Similarly if you had three pots, coffee, tea, and orange juice, you would need two tastes before you could conclude what was in all three pots. What is important here is that you *do not* need to taste all of the pots before you are certain what is in each pot.

These ideas are used in the measuring of variables. In both examples, all the pots represent your sample, and your tasting represents your sampling. Each 'pot-tasting' represents another procedure to establish certainty for your findings. Further, your number of attempts will be one less than the total number of pots. Degrees of freedom can be visualised in the same way. Whenever you use a statistical test you will always be asked to calculate your degrees of freedom. This is always very straightforward and easy to do, and does not differ for each test.

But you will note that with the measurement of any variable the degrees of freedom for that variable is the size of the sample minus either 1 or 2. We will illustrate why it is either 1 or 2 as we encounter each test. However, this 1 or 2 is best understood within the pot example above, illustrating that you don't need all the information.

The second element is the incorporation of these statistics into the text, to help the reader understand and conceptualise your findings. In writing the text, use the Describe and Decide rule to inform your reader of your finding:

- Remind the reader of the two variables you are examining.
- Describe the relationship between the two variables as positive or negative.
- Tell the reader whether the finding is significant or not.

You can use all the information above to write a fairly simple sentence, which conveys your findings succinctly, but effectively. Therefore, using the findings above, we might report:

A Pearson product–moment correlation coefficient was used to examine the relationship between scores on the Belief in Good Luck Scale and scores on the Depression–Happiness Scale. A significant positive correlation was found between Belief in Good Luck and Happiness (r[58] = 0.403, p < 0.01).

Writing a non-significant result for Pearson product–moment correlation coefficient

Knowing how to write a significant result is of little use to you if, in your research, you have a non-significant finding. Therefore, whenever we suggest how you might write test results, we will also give you the alternative way of writing up your results.

A Pearson product–moment correlation coefficient was used to examine the relationship between scores on the Belief in Good Luck Scale and scores on the Depression–Happiness Scale. No significant correlation was found between Belief in Good Luck and Happiness (r[58] = 0.043, p > 0.05).

You will of course come across different ways of writing up the Pearson product–moment correlation coefficient, both in your own writing and that of other authors, but you will find all the information included above in any write-up.

Discussing the Pearson product–moment correlation

There are three important factors to consider when you are discussing your findings.

1. *Correlation does not represent causation.* It is important to remember when reporting any sort of correlation, not to immediately infer that one variable *causes* another. Therefore, in the example above we could not conclude that Belief in Good Luck causes Happiness, as it may be that people who are happy adopt Belief in Good Luck attitudes. It is more likely that the two variables influence each other, and/or certainly work together. Remember, in your wording reflect this and always talk about relationships, associations, correlations between two variables, and not that one variable causes another, unless you have a very good reason for thinking so.

2. *Size of correlation.* This is an important consideration, and is sometimes necessary to highlight. Sometimes researchers, when their findings are significant, highlight the size of the correlation (the *r* statistic), to consider the weight of their findings. Cohen (1969) provides some well used guidelines, with correlation statistics around 0.3 and below viewed as small, 0.5 as medium and 0.7 as large. Again, remember these are used as indicators of your findings to help you in your consideration. If you have a significant correlation of 0.25, it is perhaps important not to conclude that this is a strong relationship. More importantly, you may have found a number of significant correlations in your study. Cohen's criteria can be used to determine which are more important findings, and which are less important.

3. *Association.* You will often find authors reporting the association between the two variables in a Pearson product–moment correlation coefficient, and this is thought to represent the shared variance between two variables. In theory, two variables can share a maximum of 100 per cent of the variance (identical) and a minimum of 0 per cent of variance (not related at all), and the association can be used to indicate the importance of a significant relationship between two variables. The association is found by squaring (multiplying by itself) the *r* value, and reporting it as a percentage. In the example above, the correlation between Belief in Good Luck and Happiness is $0.403 \times 0.403 = 0.1624$, and reported as a percentage, 16.24 per cent. You often find researchers reporting the variance as part of their discussion, sometimes used as an indicator of importance of the findings (or lack of importance of findings – as the smaller the percentage the less important any relationship between the two variables).

Exercise 2: Self-assessment of the Pearson product–moment correlation coefficient statistic

Read the Day *et al.* (1999) article again. Determine the following with regard to the relationship between Belief in Good Luck and Anxiety.

1. What is the *r* value for the relationship between Belief in Good Luck and Anxiety?

2. Day *et al.* (1999) do not report their degrees of freedom. What would the degrees of freedom be, for the present sample, for the relationship between Belief in Good Luck and Anxiety?

3. What is the direction of the relationship between Belief in Good Luck and Anxiety?

4. Is the relationship between Belief in Good Luck and Anxiety significant?

Figure 5.5   Computing the Pearson product–moment correlation by hand calculation.

The following example reflects a study that tried to examine the relationship between Belief in Good Luck and Happiness. The study used established scales among a non-clinical sample of five respondents. The aim of the study was to see whether higher levels of happiness accompanied a higher level of Belief in Good Luck.

Respondent	Belief in Good Luck score	Happiness score	Belief in Good Luck squared	Happiness squared	Belief in Good Luck by Happiness
1	5	4	25 (Step 1)	16 (Step 2)	20 (Step 3)
2	1	2	1 (Step 1)	4 (Step 2)	2 (Step 3)
3	1	1	1 (Step 1)	1 (Step 2)	1 (Step 3)
4	5	5	25 (Step 1)	25 (Step 2)	25 (Step 3)
5	3	3	9 (Step 1)	9 (Step 2)	9 (Step 3)

Step 1: Square each of the scores for the Belief in Good Luck variable:
$5 \times 5 = 25$, $1 \times 1 = 1$, $1 \times 1 = 1$, $5 \times 5 = 25$, $3 \times 3 = 9$

Step 2: Square each of the scores for the Happiness variable:
$4 \times 4 = 16$, $2 \times 2 = 4$, $1 \times 1 = 1$, $5 \times 5 = 25$, $3 \times 3 = 9$

Step 3: Multiply each of the scores for the Belief in Good Luck variable by each of the scores for the Happiness variable:
$5 \times 4 = 20$, $1 \times 2 = 2$, $1 \times 1 = 1$, $5 \times 5 = 25$, $3 \times 3 = 9$

Step 4: Add all the scores together for the Belief in Good Luck variable:
$5 + 1 + 1 + 5 + 3 = 15$

Step 5: Add all the scores together for the Happiness variable:
$4 + 2 + 1 + 5 + 3 = 15$

continued

Figure 5.5   (*continued*)

Step 6:   Add all the scores together for the Belief in Good Luck variable squared (from Step 1):
$25 + 1 + 1 + 25 + 9 = 61$

Step 7:   Add all the scores together for the Happiness squared (from Step 2):
$16 + 4 + 1 + 25 + 9 = 55$

Step 8:   Add all the scores together for the Belief in Good Luck variable multiplied by Happiness variable (from Step 3):
$20 + 2 + 1 + 25 + 9 = 57$

Step 9:   Multiply your finding from Step 8 by number of people in the sample:
$57 \times 5 = 285$

Step 10:   Multiply your finding from Step 4 by your finding for Step 5:
$15 \times 15 = 225$

Step 11:   Subtract your finding from Step 10 from the finding for Step 9:
$285 - 225 = 60.$

Step 12:   Multiply your finding for Step 6 by the number of people in the sample:
$61 \times 5 = 305$

Step 13:   Square your finding for Step 4:
$15 \times 15 = 225$

Step 14:   Subtract your finding for Step 13 from your finding for Step 12:
$305 - 225 = 80$

Step 15:   Multiply your finding for Step 7 by the number of people in the sample:
$55 \times 5 = 275$

Step 16:   Square your finding for Step 5:
$15 \times 15 = 225$

Step 17:   Subtract your finding for Step 16 from your finding for Step 15:
$275 - 225 = 50$

Step 18:   Multiply your finding for Step 14 by the finding for Step 17:
$80 \times 50 = 4000$

Step 19:   Find the square root for Step 18:
Sqrt $4000 = 63.245$

Step 20:   Divide the finding for Step 11 by the finding for Step 19:
$60/63.245 = 0.949$

You should now have the value of $r = 0.949$

Now you have to determine whether the figure is significant. Here, the procedure, for all hand calculations, is very different from using SPSS for Windows. You need to establish the significance level by comparing your test statistic with a pre-determined number, to determine whether your result is significant or not (this is not as exact as doing it on the computer, because, remember, this was the system used before such calculations were available on computer).

The first step is to determine the degrees of freedom. Degrees of freedom (df) for Pearson product–moment correlation coefficient is the size of sample −2 (with the minus 2 representing 1 for each set of scores). So in this case, $5 - 2 = 3$, df = 3. You then determine the significance level for your test and choose either

continued

Figure 5.5   (*continued*)

a one-tailed or two-tailed test and compare your *r* value with numbers in the table below.

We will use the $p = 0.05$ criterion and use a one-tailed statistic because the research suggests that Belief in Good Luck is accompanied by Happiness. The number that falls under $df = 3$, $p = 0.05$ and use of two-tailed test is 0.8054. If your *r* figure is higher (ignore whether the figure is + or –) than your df figure then the relationship is significant. On this occasion 0.949 (*r* figure) is higher than 0.8054 (df figure). Therefore, Pearson's *r* is significant here.

You now know your correlation statistic (*r*) and whether your finding is significant or not, and at what level it is significant. You then write the Pearson product–moment correlation in the same way.

You must also remember not to mix up the procedure for determining significance with the one for interpreting significance. Do not mix up the SPSS way and the handwritten way of doing it, or else you will make mistakes.

	Significance levels for 2–tailed test				
	0.10	0.05	0.02	0.01	0.001
	Significance levels for 1–tailed test				
$df = n - 2$	0.05	0.025	0.01	0.005	0.0005
1	0.98769	0.99692	0.999507	0.999877	0.9999988
2	0.90000	0.95000	0.98000	0.990000	0.99900
3	0.8054	0.8783	0.93433	0.95873	0.99116
4	0.7293	0.8114	0.8822	0.91720	0.97406
5	0.6694	0.7545	0.8329	0.8745	0.95074

Parametric test 2: Related *t*-test (paired-samples *t*-test in SPSS for Windows)

The related *t*-test is used to examine the differences between scores on two continuous variables that you wish to treat as parametric data. The important difference between this test and the Pearson product–moment correlation coefficient is that, while the Pearson test establishes whether two different variables are correlated, the related *t*-test seeks to establish whether scores on the same measure, administered to the same sample on two occasions, differ significantly. The related *t*-test does this by comparing the average mean scores of the same subjects in two conditions, or at two points in time.

For example, a researcher may wish to examine students' anxiety levels before and after an exam. The researcher will administer, to each student, a measure of anxiety before the exam (where the researcher might suspect anxiety levels might be high), and a measure of anxiety after the exam (where anxiety levels might be expected to be lower), to examine whether a *significant* difference occurs between anxiety levels before and after exams.

Different names for the related *t*-test

The related *t*-test also goes under a number of different names. Some-times the test is referred to as the related samples *t*-test, paired test, dependent groups *t*-test, *t*-test for related measures, correlated *t*-test, matched groups *t*-test, and most importantly for our needs, the paired-samples *t*-test in SPSS for Windows. Don't worry. These are just differ-ent terms for the same test. However, one point to note, is that the last 'matched groups *t*-test', is so named because the test is sometimes used under other circumstances. Sometimes researchers cannot administer the same measure twice to the same sample, and instead have to use two samples. Here, researchers will try to match their sample in as many ways as possible (it may be by variables such as sex, age, educa-tional attainment, length in treatment, etc.) to simulate using the same sample. On these occasions you will find researchers using a related *t*-test, referring to it as a matched groups sample.

The related *t*-test in SPSS for Windows

For this exercise we are going to concentrate on two administered measures of Neuroticism, NEUROT and NEUROT2. The scale is from the Abbreviated Form of the Revised Eysenck Personality Questionnaire (Francis *et al.*, 1992) and contains a six-item measure of neuroticism, which indicates feelings of anxiety, worry, moodiness, and frequently depression. Example items include 'Does your mood often go up and down?', 'Would you call yourself a nervous person?' and 'Do you suffer from nerves?'. Respondents are asked to rate each item on a two-point scale: Yes (scored as 1), and No (scored as 0). Possible scores range from 0 to 6 with higher scores indicating a higher level of neuroticism.

Neuroticism is a personality variable, and one of the main aspects of personality is that personality is thought to be fairly stable over time. That is, if a researcher was to measure Neuroticism with the same measure on two occasions it would be expected that individual scores would be largely similar. In this example, we have the following information among our sample. The first variable, NEUROT, is the Neuroticism scale filled out by 60 respondents during the administration of the rest of the questionnaire. NEUROT2 is the same scale used to measure Neuroticism administered to the same respondents six months later. We can use this data to examine the idea of whether the Neuroticism personality variable is stable over time. To this end, we would expect that there would be no significant difference between mean scores on the neuroticism measures.

Load up the dataset. You select the related *t*-test by clicking on the **Analyse** pull-down menu, then **Compare Means** and then **Paired-Samples**

T Test... (the name SPSS for Windows uses for the related *t*-test). Highlight the two variable names 'NEUROT' and 'NEUROT2' in the left hand box (by clicking on them with the mouse). You will see that both of the variables appear in the Current Selection Box, next to Variable 1: and Variable 2. Then click on the > box to transfer the two variables into the **Paired Variables:** box (see Figure 5.6).

One- and two-tailed hypothesis and significance testing

In the last test (the Pearson product–moment correlation) we were able to make a distinction between one-tailed and two-tailed tests. However, in SPSS there is no option for this. Nevertheless, one-tailed and two-tailed testing can be used in these sorts of tests. For the present test, in which we are comparing means scores for a continuous variable on two occasions, a one-tailed hypothesis would state that the mean score on one occasion would be significantly higher (or lower) than the mean score on another occasion. A two-tailed hypothesis would not make a statement regarding the direction of the relationship (it would not state whether mean scores would be higher or lower, there would just be a statement regarding a significant difference).

Therefore, you may be making a one-tailed hypothesis for the test, and SPSS for Windows does not provide an option for this, as its default setting is a two-tailed test. However, with the advent of computer statistics software you can easily work out the one-tailed probability statistic from the two-tailed probability statistic, by halving the two-tailed probability statistic. So, for example, if you had a two-tailed probability value of 0.50, your one-tailed probability would be 0.25.

A further point emerges from this calculation. If a researcher found a two-tailed probability value of $p = 0.08$ (not significant at the 0.05 criteria level) in determining differences between two mean scores, the one-tailed probability value of $p = 0.04$ would be significant at the 0.05 criteria level. Therefore, it seems that making one-tailed hypotheses is advantageous, as there is a better chance of there being a significant finding. As a response to the fact that finding a significant relationship may depend on how you word your hypothesis, there is a growing trend among researchers to err on the side of caution, and use two-tailed significance testing regardless of the original hypothesis (therefore, don't be surprised to find this happening when reading the research literature), because they are concerned with making a Type I error (see Chapter 4), suggesting there is a significant relationship or difference between two variables when in fact there is no significant relationship. Of course this should not detract from the fact that many researchers have very good reasons for stating one-tailed hypotheses and will continue to use one-tailed hypothesis testing.

Figure 5.6 Related (paired-samples) *t*-test window.

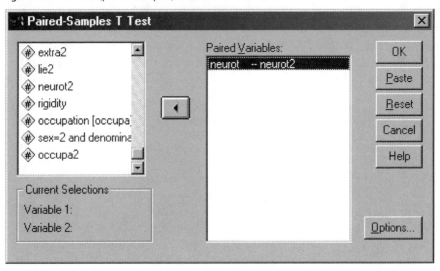

Click on **OK** to run the *t*-test. The output should look like Table 5.2.

In this output, we have all the information we need to interpret whether a significant difference occurs between scores for the two administrations of the neuroticism scale. Again, remember to *Describe* and then *Decide*.

Table 5.2 Related (paired-samples) *t*-test output.

Paired-samples statistics				
	Mean	N	Std. deviation	Std. error mean
Pair NEUROT	2.5167	60	0.9999	0.1291
1 NEUROT2	2.5000	60	0.8925	0.1152

	Paired-samples test							
	Paired differences							
				95% confidence interval of the difference				
	Mean	Std. deviation	Std. error mean	Lower	Upper	t	df	Sig. (2-tailed)
Pair 1 NEUROT – NEUROT2	1.667E–02	1.1570	0.1494	−0.2822	0.3156	0.112	59	0.912

Using Describe and Decide to interpret the related *t*-test

From the output, you will need to consider three things:

- *Mean scores and standard deviations.* These are the basis of our description. We note both the mean scores (with the standard deviation) and note which mean score for which administration is higher.
- *The t value.* The statistical test statistic. Unlike the Pearson product–moment correlation, it is not important to note whether the statistic is positive or minus.
- *The Sig. (2-tailed).* The significance level. This is the probability level given to the current findings. The significance level found in bullet point 2 tells the researcher whether the difference noted between the means (here that respondents score lower on the second administration of Neuroticism) is different. Remember, if this figure is below the $p = 0.05$ or $p = 0.01$ criteria then the finding is significant. If this figure is above 0.05 then the findings are not significant.

Therefore, the average mean score for Neuroticism on the first administration is 2.52 (with a standard deviation of 1.00) and the average mean score for Neuroticism on the second administration is 2.50 (with a standard deviation of 0.89) (note all these figures are rounded to two decimal places). We note that scores for the second administration are lower than for the first administration. The *t* value for the statistic is 0.112. This tells us very little at this stage. However, the significance level is $p = 0.912$. This is greater than 0.05. Therefore, we conclude that there is not a significant difference for mean scores between the two administrations of Neuroticism. This suggests, among our sample, that respondents do not differ significantly in their Neuroticism scores over a six-month period.

Using Decide and Describe to report the related *t*-test

The next stage is that you will need to report these statistics. There is a formal way of reporting the related *t*-test. This comprises two elements. First there is a formal statement of your statistics, and must include:

- *The test statistic.* Each test has a symbol for its statistic. The related *t*-test has the symbol *t*. Therefore, in your write-up you must include what *t* equals. In the example $t = 0.112$.
- *The degrees of freedom.* This was introduced in the last chapter, and is traditionally reported (though it is worth noting that it is not always reported). For the related *t*-test, the degrees of freedom equal the size of your sample minus 1. Here, the minus 1 represents minus 1 for the sample, because you have only asked one set of respondents. This is

placed between the *t* and the = sign and placed in brackets. Here, the degrees of freedom are 59 (size of sample = 60, minus 1 = 59). Therefore, $t(59) = 0.112$.

- *You must report the probability.* Again, this is done in relation to whether your probability value was below 0.05 or 0.01 (significant) or above 0.05 (not significant). You use less than (<) or greater than (>) the criterion level. You state this criterion by stating whether $p < 0.05$ (significant), $p < 0.01$ (significant) or $p > 0.05$ (not significant). In the example above, as $p = 0.912$, we would write $p > 0.05$ and place this after the reporting of the *t* value. Therefore, with our findings, $t(59) = 0.112$, $p > 0.05$.

This must then be incorporated into the text, to help the reader understand and conceptualise your findings. In writing the text use the Describe and Decide rule to inform your reader of your finding:

- Remind the reader of the two variables you are examining.

- Describe which mean score is highest.

- Tell the reader whether the finding is significant or not.

You can use all the information above to write a fairly simple sentence, which conveys your findings succinctly but effectively. Therefore, using the findings above we might report:

A related t-test *was used to examine differences between mean scores on the Neuroticism Scale over a six month period. No significant difference* $(t[59] = 0.112, p > 0.05)$ *was found for mean scores for Neuroticism for the first administration (mean = 2.52, SD = 1.00) and the second administration (mean = 2.50, SD = 0.89).*

Writing a significant result for related *t*-test

Example: A related *t*-test was used to examine differences between mean scores on the Neuroticism Scale over a six-month period. Scores for the first administration of Neuroticism (mean = 4.34, SD = 1.02) were found to be significantly higher $(t[59] = 5.30, p < 0.05)$ than scores for the second administration of the measure of Neuroticism (mean = 3.24, SD = 1.26).

Again, you will come across different ways of writing tests, but you will find all the information included above in any write-up.

Figure 5.7 Related *t*-test by hand calculation.

Five students are due to sit an exam. The researcher is interested in finding out whether anxiety levels change before and after an exam. The same scale is administered to the students before and after the exam, where students are asked to rate on a scale of (1) Not anxious at all, to (10) Very anxious. The following scores were obtained.

Student	Anxiety level before the exam	Anxiety level after the exam
1	7	4
2	5	6
3	6	4
4	7	3
5	8	3

Step 1: Find the mean score for the variable 'Anxiety level before the exam':

$$\frac{7 + 5 + 6 + 7 + 8}{5} = \frac{33}{5} = 6.6$$

Step 2: Find the mean score for the variable 'Anxiety level after the exam':

$$\frac{4 + 6 + 4 + 3 + 3}{5} = \frac{20}{5} = 4.0$$

Step 3: Subtract the smaller value from Steps 1 + 2 from the other:
6.6 − 4.0 = 2.6

Step 4: Subtract each number in the column 'Anxiety level after the exam' from its partner number in the column 'Anxiety level before the exam':
7 − 4 = 3, 5 − 6 = −1, 6 − 4 = 2, 7 − 3 = 4, 8 − 3 = 5.

Step 5: Then square each number calculated in Step 4, and then add them together:
3 × 3 = 9, 1 × 1 = 1, 2 × 2 = 4, 4 × 4 = 16,
5 × 5 = 25 . . . . 9 + 1 + 4 + 16 + 25 = 55

Step 6: Add up all the numbers calculated in Step 4 (take minus signs into account).
3 + −1 + 2 + 4 + 5 = 13

Anxiety level before the exam	Anxiety level after the exam	Step 4	Step 5	Step 6
7	4	3	9	3
5	6	−1	1	−1
6	4	2	4	2
7	3	4	16	4
8	3	5	25	5
			55	13

Step 7: Square the value found in Step 6: 13 × 13 = 169

Step 8: Count the number of pairs of scores: 5

Step 9: Subtract 1 from the number of pairs of scores: 5 − 1 = 4 (this also gives the degrees of freedom)

continued

Figure 5.7 (*continued*)

Step 10: Multiply the values found in Step 8 and 9: $5 \times 4 = 20$

Step 11: Divide the number found in Step 7 by Step 10: $169/20 = 8.45$

Step 12: Subtract the value found in Step 11 from that found in Step 5 and divide this number by the number found in Step 10: $(8.45 - 55/20 = -2.33)$

Step 13: Square root the value obtained in Step 12: 1.53

Step 14: Divide the value found in Step 3 by the value found in Step 13: $2.6/1.53 = 1.699$. This is your t value.

Use this table to determine the significance of your result.

	Significance levels for 2–tailed test				
	0.05	0.025	0.01	0.0005	0.00005
	Significance levels for 1–tailed test				
df	0.10	0.05	0.02	0.01	0.001
1	6.314	12.71	31.82	63.66	636.6
2	2.92	4.303	6.969	9.925	31.6
3	2.353	3.182	4.541	5.841	12.92
4	2.132	2.776	3.747	4.604	8.610
5	2.015	2.571	3.365	4.032	6.869

Degrees of freedom is what you calculated in Step 9. $= 4$

We will use a two-tailed test because the researcher has not predicted that anxiety changes in any particular direction. We will also use a significance level of 0.05. The value for degrees of freedom = 4, and using a two-tailed test, at a significance level of 0.05 the value to be compared is 2.132. t is 1.699. This is smaller than the value quoted in the table and, therefore, we would conclude there is no significant difference in anxiety levels for students before and after the exam.

Parametric test 3: Independent-samples t-test

The independent-samples t-test is used to compare mean scores for a continuous variable (that you want to treat as parametric data) by two levels of a categorical variable. A common example would be to compare males and females (a two-level categorical variable) for scores on a scale to examine for sex differences in a construct. Similar to the related t-test, the independent-samples t-test allows the researcher to compare means, but with the independent-samples t-test the research question centres on whether there is a significant difference *between* groups of respondents.

Different names for the same independent-samples *t*-test

As with the related *t*-test, this test is known under a number of names. These include 'unmatched *t*-test', '*t*-test for two independent means', 'independent *t*-test', '*t*-test for unrelated samples', and 'Student's *t*-test'. Again, do not concern yourself, they all refer to the same test.

Let us consider the Day *et al.* (1999) Belief in Good Luck and General Health paper. Within this paper there is no examination of how males and females may differ in their scores on Good Luck. Similarly, within this chapter we have used two other scales, Happiness and Neuroticism. Rather than just concentrating on one variable, let's take some aspects of this research further to examine whether males and females differ in their scores on these three scales.

Independent-samples *t*-test using SPSS for Windows

To perform the independent-samples *t*-test using SPSS for Windows, click on the **Analyse** pull-down menu and click on **Compare Means**, then click on **Independent-Samples T Test**. You will then get a screen that looks like Figure 5.8.

Highlight the variables Belief in Good Luck [BIGL], Happiness [DEPP-HAPP] and the Neuroticism [NEUROT] measure in the left hand box and move them into the **Test Variables[s]:** box by clicking on the >. Similarly, highlight the grouping variable Sex and transfer it to the **Grouping Variable**. At this point the **Grouping Variable** box will appear with [??]. You now have to assign the values of the categorical variable by clicking on

Figure 5.8   Independent-samples *t*-test.

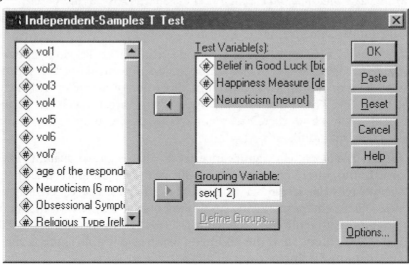

Figure 5.9   Define Groups window.

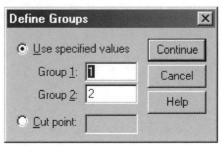

Define Groups (Figure 5.9). Now type the values of our categorical variable into the boxes. For sex the values are '1' for males, '2' for females. Type the value 1 into **Group 1** box and type 2 into the **Group 2** box, then click on **Continue**. The values 1,2 will appear in the brackets after Sex.

Now click on **OK** to run the *t*-test. Our output should look like Table 5.3.

In this output we have all the information we need to interpret whether a significant difference occurs between males and females for their scores on the Belief in Good Luck, Happiness and Neuroticism scales. Again, remember to *Describe* and then *Decide*. For each scale you must interpret the findings separately, going through each scale one at a time. Let us start with the findings for Belief in Good Luck. Please note that this procedure and write-up is almost identical to the procedure for the related *t*-test.

Using Describe and Decide to interpret the independent-samples *t*-test

From the output, you will need to consider three things:

- *Mean scores and standard deviations.* These are the basis of our description. We note both the mean scores (with the standard deviation) and note which mean score is higher.

- *The t value.* The statistical test statistic. Unlike the Pearson product–moment correlation, it is not important to note whether the statistic is positive or minus. You will notice that you are provided with two *t* statistics. Read the *Equal variances assumed* line, because you are using a parametric test and, therefore, you have assumed you are using continuous data that fulfils this criterion.

- *The Sig. (2-tailed).* The significance level. This is the probability level given to the current findings. The significance level found in bullet-point 2 above tells the researcher whether the difference noted between the means (here, that respondents score lower on the second administration of Neuroticism) is different. Remember, if this figure is below the p = 0.05 or p = 0.01 criteria then the finding is significant. If this figure is above 0.05 then the findings are not significant.

Table 5.3   Independent-samples *t*-test output.

Group statistics					
	Sex of the respondent	N	Mean	Std. deviation	Std. error mean
Belief in Good Luck	Male	29	30.3448	7.0167	1.3030
	Female	31	38.8387	7.2576	1.3035
Happiness Measure	Male	29	45.2414	8.4543	1.5699
	Female	31	47.7419	7.1273	1.2801
Neuroticism	Male	29	2.5862	1.1501	0.2136
	Female	31	2.4516	0.8500	0.1527

Independent-samples test										
		Levene's test for equality of variances		*t*-test for equality of means					95% confidence interval of the difference	
		F	Sig.	t	df	Sig. (2-tailed)	Mean difference	Std. error difference	Lower	Upper
Belief in Good Luck	Equal variances assumed	0.043	0.837	−4.603	58	0.000	−8.4939	1.8452	−12.1874	−4.8004
	Equal variances not assumed			−4.609	57.933	0.000	−8.4939	1.8431	−12.1833	−4.8045
Happiness Measure	Equal variances assumed	1.529	0.221	−1.242	58	0.219	−2.5006	2.0141	−6.5322	1.5311
	Equal variances not assumed			−1.234	54.942	0.222	−2.5006	2.0257	−6.5602	1.5591
Neuroticism	Equal variances assumed	2.484	0.120	0.518	58	0.607	0.1346	0.2599	−0.3857	0.6549
	Equal variances not assumed			0.513	51.402	0.610	0.1346	0.2625	−0.3923	0.6615

Therefore, the average mean score for males for the Belief in Good Luck Scale is 30.34 (with a standard deviation of 7.02) and the average mean score for females on the Belief in Good Luck Scale is 38.84 (with a standard deviation of 7.25). Here, we note that female scores are higher than males on Belief in Good Luck. The *t* value for the statistic is −4.603. This tells us very little at this stage. However, the significance level is p = 0.000. This is smaller than 0.01. Therefore, we conclude that there *is* a significant difference for mean scores of Belief in Good Luck between males and females. This finding suggests that females score significantly higher than males on the Belief in Good Luck Scale.

Levene's test for equality of variance

This issue reinforces our learning with regard to treating continuous data as parametric or non-parametric. The criterion set down for using an independent-samples *t*-test is that the standard deviations (variances) of the continuous data are similar. As mentioned in Chapter 5, other researchers will not note this criterion, as they will assume that the continuous variable is derived from a scale that is well established, as we did above. What Levene's test for equality of variance allows you to do, is not to worry about this debate (in sympathy with the ethos of this whole book), and gives you the option to use an independent-samples *t*-test even when your standard deviations (variances) are not similar (this sort of statistic is known as a correction statistic, because you are seeking to correct possible errors).

You will see in the second and third column of Table 5.3 the statistics for Levene's test for equality of variances, with the test statistic (F) and then the significance value. If this significance value is greater than 0.05 you can assume equal variances. If this happens, as in this case, it is recommended that you just ignore this statistic.

However, if it is significant, do not worry. Just read the *Equal variances not assumed* line. You are also recommended to report this when you write up your analysis. If you do this you need to include a simple line such as 'Levene's test for equality of variance suggested that equal variances could not be assumed (F = test statistic, $p < 0.05$), therefore *t* was corrected for equal variances not assumed.

Using Decide and Describe to report the independent-samples *t*-test

The next stage is that you will need to report these statistics. There is a formal way of reporting the independent-samples *t*-test. This comprises two elements. First there is a formal statement of your statistics, which must include:

- The test statistic. Each test has a symbol for its statistic. The independent-samples *t*-test uses the symbol *t*. Your write-up must include what *t* equals. In the example above $t = -4.603$.

- The degrees of freedom. This was introduced in the last chapter, and is traditionally reported (though it is worth noting that it is not always reported). For the independent-samples *t*-test the degrees of freedom equal the size of your sample minus 2. Here the minus 2 represents minus 1 for each sample, because you have only asked two sets of respondents (males and females). This is placed between the *t* and the = sign and placed in brackets. Here the degrees of freedom are 58 (size of sample = 60, minus 2 = 58). Therefore, $t(58) = -4.603$.

- You must report the probability. Again, this is done in relation to whether your probability value was below 0.05 or 0.01 (significant) or above 0.05 (not significant). Here, you use less than (<) or greater than (>) the criteria level. You state this criterion by stating whether $p < 0.05$ (significant), $p < 0.01$ (significant) or $p > 0.05$ (not significant). So in the example above, as $p = 0.000$, we would write $p < 0.01$ and place this after the reporting of the t value. Therefore, with our findings, $t(58) = -4.603$, $p < 0.01$.

This must then be incorporated into the text, to help the reader understand and conceptualise your findings. In writing the text use the Describe and Decide rule to inform your reader of your finding:

- Remind the reader of the two variables you are examining.
- Describe which mean score is highest.
- Tell the reader whether the finding is significant or not.

You can use all the information above to write a fairly simple sentence, which conveys your findings succinctly but effectively. Therefore using the findings above we might report:

An independent-samples t-test was used to examine differences between males and females for their mean scores on the Belief in Good Luck Scale. Females (mean = 30.34, SD = 7.02) scored significantly higher (t[58] = -4.603, p < 0.01) than males (mean = 38.84, SD = 7.26) on the measure of Belief in Good Luck.

Writing a non-significant result for the independent-samples t-test

Example: An independent-samples t-test was used to examine differences between males and females for their mean scores on the Belief in Good Luck Scale. No significant difference ($t[58] = -1.203$, $p > 0.01$) occurred between females (mean = 30.34, SD = 7.02) and males (mean = 30.84, SD = 7.26) for scores on the measure of Belief in Good Luck.

Again, you will come across different ways of writing tests, but you will find all the information included above in any write-up.

Exercise 3: Self-assessment for the independent-samples t-test

Using the datafile, examine whether a significant difference occurs between males and females for their scores on the other two measures used in the last analysis, Happiness and Neuroticism. For each, write a full paragraph, reporting the findings.

Figure 5.10 Calculation of independent-samples *t*-test by hand.

Neuroticism scores are compared by males and females, to see if there are sex differences in neuroticism. The same scale is administered to five males and five females. The researcher predicts that women will score significantly higher than men.

Men	Neuroticism score		Women	Neuroticism score	
1	4	16 (Step 1)	1	3	9 (Step 2)
2	5	25 (Step 1)	2	2	4 (Step 2)
3	5	25 (Step 1)	3	4	16 (Step 1)
4	2	4 (Step 1)	4	5	25 (Step 1)
5	3	9 (Step 1)	5	4	16 (Step 1)

First work the mean and standard deviation for men and women for their scores on Neuroticism.

Step 1: Square each of the scores for men:
$4 \times 4 = 16, 5 \times 5 = 25, 5 \times 5 = 25, 2 \times 2 = 4, 3 \times 3 = 9$

Step 2: Add up all the scores for men:
$4 + 5 + 5 + 2 + 3 = 19$

Step 3: Add up all the scores obtained from Step 1:
$16 + 25 + 25 + 4 + 9 = 79$

Step 4: Square the result of Step 2:
$19 \times 19 = 361$

Step 5: Divide the result of Step 4 by the number of men:
$361/5 = 72.2$

Step 6: Subtract the findings in Step 5 from the result in Step 3:
$79 - 72.2 = 6.8$

Step 7: Square each of the scores for women:
$3 \times 3 = 9, 2 \times 2 = 4, 4 \times 4 = 16, 5 \times 5 = 25, 4 \times 4 = 16$

Step 8: Add up all the scores for women:
$3 + 2 + 4 + 5 + 4 = 18$

Step 9: Add up all the scores obtained from Step 7:
$9 + 4 + 16 + 25 + 16 = 70$

Step 10: Square the result of Step 8:
$18 \times 18 = 324$

Step 11: Divide the result of Step 10 by the number of women:
$324/5 = 64.8$

Step 12: Subtract the findings in Step 11 from the result in Step 9:
$70 - 64.2 = 5.8$

Step 13: Add the results of Step 6 and Step 12:
$6.8 + 5.8 = 12.6$

continued

Figure 5.10   (*continued*)

Step 14: Add the number of men to the number of women and then subtract 2:
$5 + 5 - 2 = 8$

Step 15: Add the number of men to the number of women:
$5 + 5 = 10$

Step 16: Times number of men by the number of women:
$5 \times 5 = 25$

Step 17: Divide the findings of Step 15 by the findings of Step 16:
$10/25 = 0.4$

Step 18: Divide the result of Step 13 by the finding of Step 14:
$12.6/(10 - 2)$ ....$12.6/8 = 1.575$

Step 19: Multiply the result of Step 18 by the finding of Step 17:
$1.575 \times 0.4 = 0.63$

Step 20: Find the square root of Step 19:
Square root of $0.63 = 0.7937$

Step 21: Work out the Neuroticism mean score for men:
$19/5 = 3.8$

Step 22: Work out the Neuroticism mean score for women:
$18/5 = 3.6$

Step 23: Subtract the findings of Step 22 from Step 21:
$3.8 - 3.6 = 0.2$

Step 24: Divide the result of Step 23 by the finding in Step 20:
$0.2/0.7939 = 0.2519$
Your *t* value $= 0.2519$

Use this table to determine the significance of your result.

df	Significance levels for 2-tailed test			
	0.05	0.025	0.01	0.0005
	Significance levels for 1-tailed test			
	0.10	0.05	0.02	0.01
6	1.943	2.447	3.143	3.707
7	1.895	2.365	2.998	3.499
8	1.860	2.306	2.896	3.355
9	1.833	2.262	2.821	3.250
10	1.812	2.228	2.764	3.169

Degrees of freedom is what you calculated in Step 14 = 8

We will use a one-tailed test (as we are predicting women will score higher than men on Neuroticism) and we will also use a significance level of 0.05. The value for degrees of freedom = 8, one-tailed test and a significance level of 0.05 is 2.306. *t* is 0.2519. This is smaller than the value quoted in the table and, therefore, we would conclude there is no significant difference in neuroticism between men and women.

Parametric test 4: Analysis of variance (ANOVA) – between subjects

The analysis of variance works in a similar way to the independent-samples *t*-test. The simplest way to view analysis of variance, is that it is used when you want to compare scores on a continuous variable, that you are treating as parametric data, by the levels of a categorical variable with three levels or more (remember, an independent-samples *t*-test was used when the categorical variable had two levels). Apart from that, it works in a similar way to the independent-samples *t*-test, comparing the average mean score on the continuous variables by the levels of the categorical variable.

Read the paper by Maltby *et al.* (1994) given in Figure 5.11.

Figure 5.11   'Religiosity and Obsessionality'.

The Journal of Psychology, 128, (5), 609–611

Religiosity and Obsessionality: A Refinement

JOHN MALTBY
PADDY MCCOLLAM
DAVID MILLAR
Department Of Psychology
University of Ulster at Coleraine, Northern Ireland

LEWIS (1994) examined Freud's (1907/1961) observation of the resemblance between obsessive actions and religious practices by using a measure of obsessional symptoms (Sandler & Hazari, 1960) and the adult form of the Francis Scale of Attitudes towards Christianity (ASC4B; Francis & Stubbs, 1987) and found little support for Freud's observations. Pearson correlations between the two measures were, for men, $r = .19$, $p > .05$, and, for women, $r = .21$, $p > .05$.

Freud's (1907/1961) description is related to communal religious practice, and the ASC4B (used by Lewis) does not discriminate between communal worship and other forms of religious practice. Our aim in this study was to examine whether obsessional symptoms would be significantly higher in those who participate in communal worship than in those who are not religious.

The 'Age Universal' I-E scale (Gorsuch & Venable, 1983), derived from the Religious Orientation Scale (Allport & Ross, 1967), the ASC4B (Francis & Stubbs, 1987), and the Sandler-Hazari Obsessionality Inventory (Sandler & Hazari, 1960), were administered to 119 residents in the United Kingdom (85 women and 34 men). The I-E scale makes a distinction between extrinsic (i.e., communal worship) and intrinsic (other) religious practices. The scale assumes that the respondent is religious.

If our respondents indicated that the I-E scale was irrelevant to them, they were placed in a new group of nonreligious persons. This placement was crosschecked using respondents' scores on the ASC4B (Francis & Stubbs, 1987). The nonreligious

continued

Figure 5.11   (*continued*)

group had a mean score on the ASC4B of 47.7 (*SD* = 12.4), whereas the religious group had a mean of 89.5 (*SD* = 15.5).

Scores on the I-E scale located above the median were used to identify intrinsic persons, whereas scores below the median identified extrinsic persons. Though Freud's observations do not extend to intrinsic religious practices, the intrinsic group was included to complete the analysis.

Obsessional symptom scores were calculated for the extrinsic group (n = 33; *M* = 9.2, *SD* = 3.3), the intrinsic group (n = 33; *M* = 8.4, *SD* = 4), and the nonreligious group (n = 53; *M* = 7.1, *SD* = 3.7). A one-way analysis of variance (ANOVA) indicated a significant difference between the groups on the measure of obsessional symptoms, $F(2, 116) = 3.68$, $p < .05$. Comparisons indicated no significant differences between the extrinsic and intrinsic groups ($p > .05$), and the intrinsic and nonreligious groups, ($p > .05$). However, there was a significant difference between the extrinsic group and the nonreligious group, ($p < .05$). To aid in the interpretation of these results, we made a comparison between the scores on the ASC4B and the obsessional symptoms measure in accordance with Lewis (1994). Correlations in both studies were similar. Here, for $r = .21$, $p > .05$, and, for women, $r = .17$, $p > .05$.

The results suggest that communal worship is accompanied by more obsessional symptoms, as Freud claimed. However, scores did not vary much from those for intrinsic practices, which in turn varied little from obsessional symptom levels in the nonreligious group. In light of this, the empirical dismissal of Freud's (1907/1961) observations regarding the two behaviours may be premature.

REFERENCES

Allport, G.W., & Ross, J.M. (1967). Personal religious orientation and prejudice. *Journal of Personality and Social Psychology*, 5, 432–433.

Francis, L.J., & Stubbs, M.T. (1987). Measuring attitudes towards Christianity: From childhood to adulthood. *Personality and Individual Differences*, 8, 741–743.

Freud, S. (1961). Obsessive actions and religious practices. In J. Strachey (Ed. and Trans), *The standard edition of the complete psychological works of Sigmund Freud* (Vol. 9, pp. 116–127). London: Hogarth Press. (Original work published 1907)

Gorsuch, R.L., & Venable, G.D. (1983). Development of an 'Age Universal' I-E scale. *Journal for the Scientific Study of Religion*, 22(2), 181–187.

Lewis, C.A. (1994). Religiosity and obsessionality: The relationship between Freud's religious practices. *Journal of Psychology*, 128, 189–196.

Sandler, J., & Hazari, A. (1960). The obsessional: On the psychological classification of obsessional character traits and symptoms. *British Journal of Medical Psychology*, 33, 113–122.

Received November, 2, 1993

In the present study, the authors are using the distinction between three types of religiousness – extrinsic (where religion is very social-orientated), intrinsic (where religion is personal) and non-religious (where people are not religious) – to test Freud's idea that social religion resembles obsessional acts. As such, the authors place respondents into one of three categories (extrinsic, intrinsic, non-religious) and compare their levels of obsessional symptoms. By comparing the means for significant differences between groups, they find some support for Freud's ideas. The present findings are suggesting that a certain type of religious respondent (extrinsic, where religion is a social activity) is likely to show higher obsessional symptoms, than other types of religious respondents (intrinsic, where religion is very personal; and respondents who are not religious at all).

Let us see if we can replicate this finding with our dataset. In the dataset we have two variables. The first variable is RELTYPE, identification of a respondent's religious type. This is a categorical variable that indicates whether people are either extrinsically religious, intrinsically religious, or not religious (we have done the hard work and used a religiosity scale (Gorsuch and Venable, 1983) to determine these categories). In the present sample, an extrinsically religious person is scored as '3', an intrinsically religious person is scored as '2', and a non-religious person is scored as '1'. The second variable comprises scores from The Sandler–Hazari Obsessionality Inventory (Sandler and Hazari, 1960). This inventory contains a measure of obsessional symptoms. Obsessional symptoms are items describing feelings of guilt, ritualistic behaviours, indecision and compulsive thoughts. Higher scores indicate a higher level of obsessional symptoms. Therefore, the analysis of variance used here will examine whether the three categories of religious type (extrinsic/intrinsic/none) differ *significantly* from each other in their levels of obsessional symptoms. Load up the book's datafile.

Pull down the **Analyse** menu and click on **Compare Means**, and then **One-Way Anova**. You should now get a screen that looks like Figure 5.12.

Move the dependent variable Obsessional Symptoms (OBSYM) into the **Dependent List**: box. Move the variable Religious type (RELTYPE) into

Figure 5.12   One-way ANOVA window.

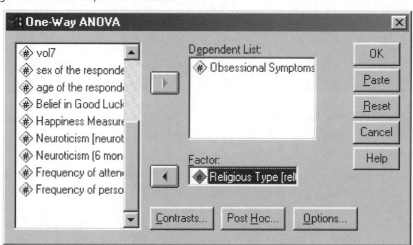

Figure 5.13   One-way ANOVA: Options.

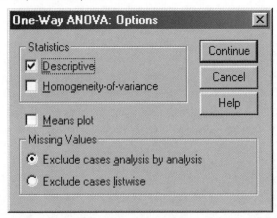

the **Factor**: box. However, before pressing **OK** click on the **Options** box (Figure 5.13).

To get the means for each religious group, we need to click on the **Descriptive** button and then click on the **Continue** box. Then click on the **Post Hoc** . . . button. You will get a new screen. Click on the box next to the **Scheffe** (we will elaborate on what this is, when we've provided the context for interpreting the test). Click on **Continue**, and then press **OK**. You will then get an output screen as in Table 5.4.

Within this output we have all the information we need to interpret whether a significant difference occurs between religious types for their scores on the obsessional symptoms scale. However, it is best to *Describe* and then *Decide* about this output in two stages.

Using Describe and Decide to interpret the analysis of variance (stage 1)

From the first two boxes in this output, you will need to consider three things:

- *Mean scores and standard deviations* (Descriptives in Table 5.4). These are the basis of our description. Here, we note both the mean scores (with the standard deviation) of each of the groups of religious persons and note which mean score is higher.
- *The F value* (ANOVA in Table 5.4) The statistical test statistic.
- *The Sig. (2-tailed)* (ANOVA in Table 5.4). The significance level. This is the probability level given to the current findings. The significance level tells the researcher whether the difference noted between the means is significantly different. Remember, if this figure is below the $p = 0.05$ or $p = 0.01$ criteria, the finding is significant. If this figure is above 0.05, the finding is not significant.

Table 5.4   Analysis of variance output.

Descriptives

Obsessional symptoms

	N	Mean	Std. deviation	Std. error	95% confidence interval for mean Lower bound	Upper bound	Minimum	Maximum
Non-religious	18	4.7778	1.9571	0.4613	3.8046	5.7510	1.00	7.00
Intrinsic	24	4.7917	1.9777	0.4037	3.9566	5.6268	1.00	8.00
Extrinsic	18	6.7222	2.4688	0.5819	5.4945	7.9499	2.00	10.00
Total	60	5.3667	2.2771	0.2940	4.7784	5.9549	1.00	10.00

ANOVA

Obsessional symptoms

	Sum of squares	df	Mean square	F	Sig.
Between groups	47.253	2	23.626	5.206	0.008
Within groups	258.681	57	4.538		
Total	305.933	59			

Multiple comparisons

Dependent variable: obsessional symptoms
Scheffe

(I) Religious type	(J) Religious type	Mean difference (I − J)	Std. error	Sig.	95% confidence interval Lower bound	Upper bound
Non-religious	Intrinsic	−1.3889E-02	0.6642	1.000	−1.6835	1.6557
	Extrinsic	−1.9444*	0.7101	0.030	−3.7293	−0.1596
Intrinsic	Non-religious	1.389E-02	0.6642	1.000	−1.6557	1.6835
	Extrinsic	−1.9306*	0.6642	0.019	−3.6001	−0.2610
Extrinsic	Non-religious	1.9444*	0.7101	0.030	0.1596	3.7293
	Intrinsic	1.9306*	0.6642	0.019	0.2610	3.6001

* The mean difference is significant at the 0.05 level.

The average mean score for respondents who are non-religious is 4.78 (with a standard deviation of 1.96), the average mean score for respondents who display intrinsic religiosity is 4.79 (with a standard deviation of 1.98) and the average mean score for respondents who display extrinsic religiosity is 6.72 (with a standard deviation of 2.47). Here, we note that extrinsic respondents score higher than both intrinsic and non-religious respondents, and that intrinsic respondents score higher than non-religious respondents. The F value for the statistic is 5.206. This tells us very little at this stage. However, the significance level is $p = 0.008$. This is smaller than 0.01, therefore, we conclude that there is a significant difference for mean scores of obsessional symptoms between religious type.

Using Describe and Decide to interpret the analysis of variance – between subjects (stage 2)

You only carry out this stage if you have found a significant result for the ANOVA. We found a significant difference between the three religious categories for scores on the obsessional symptoms; however, we do not know where the significant difference lies between the groups. Is the difference between extrinsic and intrinsic respondents, between extrinsic and non-religious respondents, or intrinsic and non-religious persons? To answer this question we use Table 5.4 (Multiple comparisons), which was generated when we pressed the post-hoc button and clicked the Scheffe test.

The Scheffe test determines significant differences between mean scores of each of the possible pairs of your categorical variable (i.e. extrinsic and intrinsic is a pairing, intrinsic and non-religious is a pairing, and extrinsic and non-religious is a pairing) to allow you an overall picture of where the differences lie. The Scheffe test provides you with a probability value to determine whether there is a significant difference between the groups. In our present output, this is presented in Table 5.4 (Multiple comparisons). In each row, each of the possible pairings are presented with each level of the categorical variable. In the fourth column of this table is the significance value that tells you whether mean scores between pairings are different. Again, if the value is less than 0.05, or 0.01, the mean scores are significantly different. If the significance value is above 0.05 it is not significant.

Using the Scheffe test

It is worth noting that there are many other post-hoc comparison tests, and that we used the Scheffe test to illustrate the analysis of variance – between subjects test. However, researchers may use different types of post-hoc tests for different reasons (for example, some researchers do not use Scheffe because it sets its criteria at a very conservative level, while other researchers will use the Scheffe test for that reason). Explaining the differences between the tests is not appropriate to the aims of this book, but SPSS for Windows provides further information on each of these tests in the Help section.

Reading down this column we can see that there is:

- In the first row (non-religious): no significant difference between non-religious respondents and intrinsic respondents (p = 1.00, p > 0.05), and a significant difference between non-religious respondents and extrinsic respondents (p = 0.03, p < 0.05).

- In the second row (intrinsic): no significant difference between intrinsic respondents and non-religious respondents (p = 1.00, p > 0.05) (it is worth noting that this information is repeating information gained from the first row), and a significant difference between intrinsic respondents and extrinsic respondents (p = 0.19, p < 0.05).

- In the third row (extrinsic): a significant difference between extrinsic respondents and non-religious respondents (p = 0.03, p < 0.05), and a significant difference between extrinsic respondents and intrinsic respondents (p = 0.19, p < 0.05) (it is worth noting that this information is repeating information from the first two rows).

Therefore, using this information, we can conclude that the significant differences occurring for the analysis of variance are the result of extrinsic religious respondents scoring significantly higher than intrinsic and non-religious respondents on the measure of obsessional symptoms.

Please remember that if you did not find a significant difference between the means in stage 1, there is no need to carry out stage 2.

Using Decide and Describe to report the analysis of variance – between subjects

The next stage is that you will need to report these statistics. There is a formal way of reporting the analysis of variance. This comprises two aspects. First there is a formal statement of your analysis of variance statistic, and must include:

- The test statistic. Each test has a symbol for its statistic. The analysis of variance uses the F statistic. Therefore, in your write-up you must include what F equals. In the example above F = 5.206.

- The degrees of freedom. Here, the degrees of freedom is slightly different from the tests you have encountered so far. Here, you report two figures. The first is the number of groups (levels of the categorical variable) minus 1; here 3 groups minus 1 = 2. The second is the number of respondents minus the number of groups; here 60 respondents minus 3 = 57. This is placed between the F and the = sign, separated by a comma and placed in brackets. Therefore, F(2, 57) = 5.206.

- You must report the probability of the analysis of variance. Again, this is done in relation to whether your probability value was below 0.05 or 0.01 (significant) or above 0.05 (not significant). Here, you use smaller than (<) or greater than (>) the criteria level. You express this criterion by stating whether p < 0.05 (significant), p < 0.01 (significant) or p > 0.05 (not significant). So in the example above, as p = 0.008, we would write p < 0.01 and place this after the reporting of the F value. Therefore with our findings, F(2, 57) = 5.206, p < 0.01.

Second, *if* you found a significant difference for the analysis of variance statistic, then you report the planned comparisons. When reporting Scheffe statistics you just state whether p was less than (<, significant) or greater than (>, not significant) 0.05.

This must then be incorporated into the text, to help the reader understand and conceptualise your findings. In writing the text use the *Describe* and *Decide* rule to inform your reader of your finding:

- Remind the reader of the two variables you are examining.
- Describe which mean score is highest.
- Tell the reader whether the analysis of variance is significant or not.
- If the analysis of variance is significant, you must inform the reader where the differences lie between groups.

You can use all the information above to write a fairly simple paragraph, which conveys your findings succinctly, but effectively. Therefore, using the findings above we might report:

Obsessional symptom scores were calculated for the extrinsic group (mean = 6.72, SD = 2.47), the intrinsic group (mean = 4.79, SD = 1.98), and the non-religious group (mean = 4.78, SD = 1.96). An analysis of variance indicated a significant difference between the groups on the measure of obsessional symptoms, F(2, 57) = 5.206, p < 0.01. A Scheffe test revealed that extrinsic religious respondents scored significantly higher than both intrinsic respondents (p < 0.05) and non-religious respondents (p < 0.05) on the measure of obsessional symptoms. However, there was no significant difference between the intrinsic respondents and the non-religious respondents (p > 0.05).

Writing a non-significant result for analysis for variance

Example: Obsessional symptom scores were calculated for the extrinsic group (mean = 4.72, SD = 2.47), the intrinsic group (mean = 4.79, SD = 1.98), and the non-religious group (mean = 4.78, SD = 1.96). An analysis of variance indicated no significant difference between the groups on the measure of obsessional symptoms, F(2, 57) = 1.206, p > 0.05.

Again, you will come across different ways of writing tests, but you will find all the information included above in any write-up.

Parametric test 5: Analysis of variance (ANOVA) – within subjects

The simplest way to view analysis of variance – within subjects, is that it is used when you want to compare scores on the same continuous variable, that you view as being able to be used in a parametric test, that have been administered on three occasions or more. The analysis of variance – within subjects works in a similar way to the related *t*-test, by comparing the same variable over a number of occasions. Further, the analysis of variance – within subjects is very similar to the analysis of variance – between subjects in that both tests use similar terms to examine the difference between administrations of the variables.

To illustrate the analysis of variance – within subjects in SPSS for Windows, let us use the Neuroticism variable that we used in the related *t*-test example. In that example, we compared mean scores of Neuroticism across two administrations (six months apart) with the view that there should be differences in scores on Neuroticism over time because it is a personality variable. However, we have a further administration of Neuroticism (Neurot3) in the dataset that represents the same sample's scores on Neuroticism a year after the original administration (Neurot) and six months after the second administration (Neurot2).

Pull down the **Analyse** menu and click on **General Linear Model**, and then on **Repeated Measures...** You should now get a screen that looks like Figure 5.14.

This window lets us define how many occasions the continuous variable was administered. In the box next to **Number of Levels** type a 3, press **Add** and then **Define**. The following Repeated Measures window given in Figure 5.15 will then appear.

Move the variable that represents the first administration of Neuroticism (Neuroticism) into the **Within Subject Variables** box. Then repeat this procedure for the second (Neurot2) and third (Neurot3) administration of Neuroticism. Your Window should look like Figure 5.16.

Figure 5.14   Repeated Measures Define Factor[s] window.

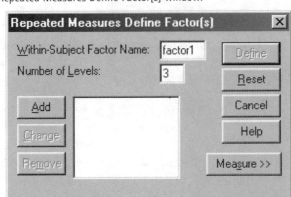

Figure 5.15 Repeated Measures window.

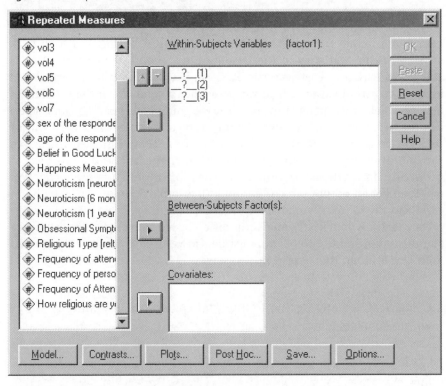

Figure 5.16 Repeated Measures window.

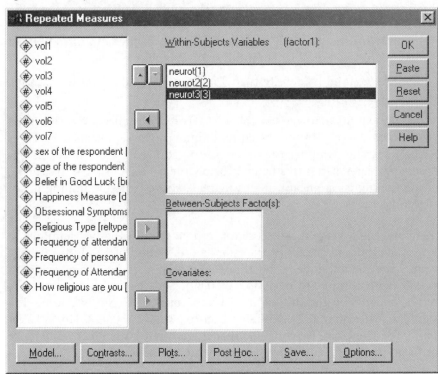

We can do two things in this window. First, to get the means for each neuroticism score, we need to click on the **Options** button and then click on **Descriptive Statistics** in the **Display** box. Second, remember with the analysis of variance – between subjects test, if a significant result was found for the analysis of variance test, we were able to break down the differences between each of the pairs of means by using the Scheffe test. We can also do this for the analysis of variance – within subjects test. In the same window click on Factor1 in the **Factor[s] and Factor Interaction** box and move Factor to the **Display Means for** using the >. Then click on the box next to **Compare Main effects**, and then click on **Continue**. Then press **OK**. You will then get an output screen that looks like Table 5.5.

Within this output we have all the information we require for an interpretation of whether a significant difference occurs between the three administrations of the Neuroticism scale. However, it is best to *Describe* and then *Decide* about this output in two stages.

Using Describe and Decide to interpret the analysis of variance – within subjects (stage 1)

From the first two boxes in this output, you will need to consider three things:

- *Mean scores and standard deviations* (Descriptive statistics in Table 5.5). These are the basis of our description. Here, we note both the mean scores (with the standard deviation) for each of the administrations of neuroticism and note which mean score is higher.

- *The F value* (Multivariate tests in Table 5.5). The statistical test statistic (here read the first statistic).

- *The Sig.* (Multivariate tests in Table 5.5). The significance level. This is the probability level given to the current findings. The significance level tells the researcher whether the difference noted between the means is different. Remember, if this figure is below the $p = 0.05$ or $p = 0.01$ criteria, the finding is significant. If this figure is above 0.05, the finding is not significant.

Therefore, the average mean score for respondents on the first administration of neuroticism is 2.52 (with a standard deviation of 1.00), the average mean score for respondents on the second administration of neuroticism is 2.50 (with a standard deviation of 0.89), and the average mean score for respondents on the third administration of neuroticism is 2.50 (with a standard deviation of 0.79). The significance level is $p = 0.992$. This is larger than 0.05. Therefore, we conclude that there is not a significant difference for mean scores of Neuroticism over the three occasions.

Table 5.5 Relevant output for analysis of variance – within subjects test.

Descriptive statistics

	Mean	Std. deviation	N
Neuroticism	2.5167	0.9999	60
Neuroticism (6 months later)	2.5000	0.8925	60
Neuroticism (1 year later)	2.5000	0.7919	60

Multivariate tests[b]

Effect		Value	F	Hypothesis df	Error df	Sig.
FACTOR1	Pillai's trace	0.000	0.008[a]	2.000	58.000	0.992
	Wilks' lambda	1.000	0.008[a]	2.000	58.000	0.992
	Hotelling's trace	0.000	0.008[a]	2.000	58.000	0.992
	Roy's largest root	0.000	0.008[a]	2.000	58.000	0.992

[a] Exact statistic
[b] Design: Intercept
  Within subjects design: FACTOR1

Pairwise comparisons

Measure: MEASURE_1

(I) FACTOR1	(J) FACTOR1	Mean difference (I − J)	Std. error	Sig.[a]	95% confidence interval for difference[a] Lower bound	Upper bound
1	2	1.667E–02	0.149	0.912	−0.282	0.316
	3	1.667E–02	0.129	0.898	−0.242	0.275
2	1	−1.667E–02	0.149	0.912	−0.316	0.282
	3	0.000	0.082	1.000	−0.165	0.165
3	1	−1.667E–02	0.129	0.898	−0.275	0.242
	2	0.000	0.082	1.000	−0.165	0.165

Based on estimated marginal means.
[a] Adjustment for multiple comparisons: least significant difference (equivalent to no adjustments).

Using Describe and Decide to interpret the analysis of variance – between subjects (stage 2)

You only do this stage if you found a significant result for the analysis of variance – within subjects. In the last example we didn't find a significant result for the tests so we would not need to use it. However, you need to know what to do if you do find a significant result for the test. To gain this information we use Table 5.5 (Pairwise comparisons).

What this table does is show the test (least significant difference) used for finding significant differences between mean scores of each of the possible pairs of administration (i.e. the first administration and second administration is a pairing, the second administration and third administration is a pairing, and the first administration and third administration is a pairing) to allow you an overall picture of where the differences lie. As with the Scheffe test in the analysis of variance – between subjects test, the least significant difference test provides a probability value to determine whether there is a significant difference between the administrations. In our present output, this is presented in Table 5.5 (Pairwise comparisons). In each row, each of the possible pairings are presented with each level of the categorical variable. In the fourth column of this table is the significance value that tells you whether mean scores between administrations are different. Again, if the value is less than 0.05, or 0.01, the mean scores are significantly different; if the significance value is above 0.05, it is not significant.

Using this table as an example, reading down this column we can see that there is:

- No significant difference between Administration time 1 and Administration time 2 ($p = 0.912$, $p > 0.05$), and no significant difference between Administration time 1 and Administration time 3 ($p = 0.898$, $p > 0.05$).

- No significant difference between Administration time 2 and Administration time 1 ($p = 0.912$, $p > 0.05$), and no significant difference between Administration time 2 and Administration time 3 ($p = 1.00$, $p > 0.05$).

- No significant difference between Administration time 3 and Administration time 1 ($p = 0.898$, $p > 0.05$), and no significant difference between Administration time 3 and Administration time 2 ($p = 1.00$, $p > 0.05$) (it is worth noting that this information is repeating information from the first two rows).

Please remember that we did not find a significant difference between the means in stage 1, therefore in the present example, there is no need for us to report this.

Using Decide and Describe to report the analysis of variance – within subjects

The next stage is that you will need to report these statistics. There is a formal way of reporting the analysis of variance – within subjects. This comprises two aspects. First there is a formal statement of your analysis of variance statistic, and must include:

- The test statistic. Each test has a symbol for its statistic. The analysis of variance uses the F statistic. Therefore, in your write-up you must include what F equals. In the example above F = 0.008.

- The degrees of freedom. Here, the degrees of freedom is slightly different than the tests you have encountered so far. Here, you report two figures; the first is the number of administrations minus 1. Here, 3 administrations minus 1 = 2. The second is the number of respondents minus the number of administrations. Here, 60 respondents minus 3 = 57. This is placed between the F and the = sign, separated by a comma and placed in brackets. Therefore, $F(2, 57) = 0.008$.

- You must report the probability of the analysis of variance. Again, this is done in relation to whether your probability value was below 0.05 or 0.01 (significant) or above 0.05 (not significant). Here, you use less than (<) or greater than (>) the criteria level. You state this criterion by stating whether $p < 0.05$ (significant), $p < 0.01$ (significant) or $p > 0.05$ (not significant). So in the example above, as $p = 0.992$, we would write $p > 0.05$ and place this after the reporting of the F value. Therefore, with our findings, $F(2, 57) = 0.008$, $p > 0.05$.

Second, *if* you found a significant difference for the analysis of variance statistic, then you report the pairwise comparisons. When reporting the least significant difference statistics you just state whether p was less than (<, significant) or greater than (>, not significant) 0.05.

This must then be incorporated into the text, to help the reader understand and conceptualise your findings. In writing the text use the *Describe* and *Decide* rule to inform your reader of your finding:

- Remind the reader of the two variables you are examining.

- Describe which mean score is highest.

- Tell the reader whether the analysis of variance is significant or not.

- If the analysis of variance is significant, you must inform the reader where the differences lie between groups.

You can use all the information above to write a fairly simple paragraph, which conveys your findings succinctly, but effectively. Therefore, using the findings above we might report:

> *Neuroticism mean scores were calculated for the 1st administration of Neuroticism (mean = 2.52, SD = 1.00), the 2nd administration of Neuroticism (mean = 2.50, SD = 0.89), and the 3rd administration of Neuroticism (mean = 2.50, SD = 0.79). An analysis of variance indicated that no significant difference occurred between scores on Neuroticism for the three administrations (F(2, 57) = 0.008, p > 0.05).*

Writing a non-significant result for analysis for variance – within subjects

Example: Neuroticism mean scores were calculated for the 1st administration of Neuroticism (mean = 4.52, SD = 1.00), the 2nd administration of Neuroticism (mean = 2.50, SD = 0.89), and the 3rd administration of Neuroticism (mean = 2.50, SD = 0.79). An analysis of variance indicated a significant difference occurred between scores on Neuroticism for the three administrations ($F(2, 57) = 5.338$, $p < 0.05$). Least significance difference tests of means revealed that respondents scored significantly higher on the 1st administration of Neuroticism than on the 2nd ($p < 0.05$) and 3rd administrations ($p < 0.05$). However, there was no significant difference between mean scores for the 2nd and 3rd administrations of Neuroticism ($p > 0.05$).

Again, you will come across different ways of writing tests, but you will find all the information included above in any write-up.

Calculation of analysis of variance statistics by hand

During this chapter you have learnt a number of new skills. At this stage we shall not ask you to learn how to do an analysis of variance statistic by hand, as it may distract you from the main aims of the book. The reason for this is two-fold. The first is that working out the analysis of variance is a long, complicated statistic to calculate, and at present, there are a number of skills to learn in the chapter that are paramount to your early success. Second, analysis of variance statistics are considered an advanced statistic, and due to the number of stages involved (e.g. using multiple comparison statistics) it is prudent for you to start using SPSS for Windows.

Exercise 4: Assessment check for analysis of variance – within subjects

Table 5.6 shows the output for an analysis of variance – within subjects that makes comparisons, among 50 adults, for three administrations of the Depression–Happiness Scale (McGreal and Joseph, 1993) over a year (initially, six months later, and one year after the initial administration). The output includes a breakdown of mean scores, the analysis of variance statistic, and the least significance difference tests. Using this information determine whether a significant difference occurs for mean scores across the three administrations.

Table 5.6 Output for self-assessment for analysis of variance – within subjects.

Descriptive statistics

	Mean	Std. deviation	N
DEPPHAP1	45.3184	7.1257	50
DEPPHAP2	43.8937	9.0082	50
DEPPHAP3	48.7835	7.2511	50

Multivariate tests[b]

Effect		Value	F	Hypothesis df	Error df	Sig.
FACTOR1	Pillai's trace	0.175	5.082[a]	2.000	48.000	0.010
	Wilks' lambda	0.825	5.082[a]	2.000	48.000	0.010
	Hotelling's trace	0.212	5.082[a]	2.000	48.000	0.010
	Roy's largest root	0.212	5.082[a]	2.000	48.000	0.010

[a] Exact statistic
[b] Design: Intercept
  Within subjects design: FACTOR1

Pairwise comparisons

Measure: MEASURE_1

(I) FACTOR1	(J) FACTOR1	Mean difference (I – J)	Std. error	Sig.[a]	95% confidence interval for difference[a]	
					Lower bound	Upper bound
1	2	1.425	1.659	0.395	−1.908	4.758
	3	−3.465*	1.514	0.026	−6.508	−0.422
2	1	−1.425	1.659	0.395	−4.758	1.908
	3	−4.890*	1.609	0.004	−8.123	−1.657
3	1	3.465*	1.514	0.026	0.422	6.508
	2	4.890*	1.609	0.004	1.657	8.123

Based on estimated marginal means.
* The mean difference is significant at the 0.05 level.
[a] Adjustment for multiple comparisons: least significant difference (equivalent to no adjustments).

References

Clegg, F. (1987). *Simple Statistics*. Cambridge: Cambridge University Press.

Cohen, J. (1969). *Statistical Power Analysis for the Behavioral Sciences*. New York: Academic Press.

Day, L., Maltby, J. and Macaskill, A. (1999). The relationship between belief in good luck and general health. *Psychological Reports*, **85**, 971–972.

Francis, L.J., Brown, L.B. and Philipchalk, R. (1992). The development of an Abbreviated Form of the Revised Eysenck Personality Questionnaire (Epqr-A) – its use among students in England, Canada, the USA and Australia. *Personality and Individual Differences*, **13**, 443–449.

Gorsuch, R.L. and Venable, G.D. (1983). Development of an 'Age Universal' I-E scale. *Journal for the Scientific Study of Religion*, **22**, 181–187.

Lewis, C.A. (2000). The Depression–Happiness Scale. In J. Maltby, C.A. Lewis and A.P. Hill (Eds), *A Handbook of Psychological Tests* (Vol. 2). Cardiff: Edwin Mellen Press.

Maltby, J., McCollam, P. and Millar, D. (1994). Religiosity and obsessionality: a refinement. *The Journal of Psychology*, **128**, 609–611.

McGreal, R. and Joseph, S. (1993). The Depression–Happiness Scale. *Psychological Reports*, **73**, 1279–1282.

Sandler, J. and Hazari, A. (1960). The obsessional: on the psychological classification of obsessional character traits and symptoms. *British Journal of Medical Psychology*, **33**, 113–122.

Answers to exercises

Exercise 1: Chapter energiser

Table 5.7 Solution to Energiser Puzzle.

First team drawn	Second team drawn	Date played	Score
Manchester Utd	Rotherham	5 Nov	2–1
Liverpool	Grimsby	12 Nov	1–1
Blackpool	Arsenal	19 Nov	1–0
Sunderland	Chelsea	26 Nov	2–0
Newcastle Utd	Sheffield Wednesday	3 Dec	1–3

Exercise 2: Self-assessment of the Pearson product–moment correlation coefficient statistic

1. $r = -0.29$.

2. $df = 60$.

3. It is a negative relationship.

4. Yes, $p < 0.05$.

Exercise 3: Self-assessment for the independent-samples *t*-test

You should get a similar output to Table 5.8.

Your answer for the Happiness measure should read something like this: 'An independent-samples *t*-test was used to examine differences between males and females for their mean scores on the Depression–Happiness measure. No significant difference $(t[58] = -1.24, p > 0.05)$ occurred between females (mean = 47.74, SD = 7.13) and males (mean = 45.24, SD = 8.45) on the measure of Happiness.'

Table 5.8   Output for self-assessment for the independent-samples t-test.

Group statistics

	Sex of the respondent	N	Mean	Std. deviation	Std. error mean
Neuroticism	Male	29	2.5862	1.1501	0.2136
	Female	31	2.4516	0.8500	0.1527
Happiness measure	Male	29	45.2414	8.4543	1.5699
	Female	31	47.7419	7.1273	1.2801

Independent-samples test

		Levene's test for equality of variances		t-test for equality of means					95% confidence interval of the difference	
		F	Sig.	t	df	Sig. (2-tailed)	Mean difference	Std. error difference	Lower	Upper
Neuroticism	Equal variances assumed	2.484	0.120	0.518	58	0.607	0.1346	0.2599	−0.3857	0.6549
	Equal variances not assumed			0.513	51.402	0.610	0.1346	0.2625	−0.3923	0.6615
Happiness measure	Equal variances assumed	1.529	0.221	−1.242	58	0.219	−2.5006	2.0141	−6.5322	1.5311
	Equal variances not assumed			−1.234	54.942	0.222	−2.5006	2.0257	−6.5602	1.5591

Your answer for the Neuroticism measure should read something like this: 'An independent-samples t-test was used to examine differences between males and females for their mean scores on a measure of Neuroticism. No significant difference ($t[58] = -0.518$, $p > 0.05$) occurred between females (mean $= 2.45$, SD $= 0.85$) and males (mean $= 2.59$, SD $= 1.15$) on the measure of Neuroticism.'

Exercise 4: Assessment check for analysis of variance – within subjects

Your answer for the comparison of the Happiness measure should read something like this: 'Mean scores were calculated for the 1st administration of Happiness (mean $= 45.32$, SD $= 7.13$), the 2nd administration of Happiness (mean $= 43.89$, SD $= 9.01$), and the 3rd administration of Happiness (mean $= 48.78$, SD $= 7.25$). An analysis of variance indicated that a significant difference occurred between scores on Happiness for the three administrations ($F(2, 47) = 5.082$, $p < 0.05$). Least significant difference tests of means revealed that respondents scored significantly higher on the 3rd administration of Neuroticism than on the 1st ($p < 0.05$) and 2nd administrations ($p < 0.01$). However, there was no significant difference between mean scores for the 1st and 2nd administrations of Neuroticism ($p > 0.05$).'

CHAPTER 6

Non-parametric tests:
examining the relationship between
sex and religion

*Each problem that I solved became a rule which served
afterwards to solve other problems.*

(*René Descartes (1596–1650)*, Discours de la Méthode, *1637*)

In this session you will learn the rationale, the procedure and the interpretation for
six non-parametric tests:

- Chi-square (for use with two categorical variables).
- Spearman's rho (for use with two continuous variables, but you have decided
  to use the variables in a non-parametric test).
- Wilcoxon sign-ranks test (for use with a continuous variable that has been
  administered on two occasions, but you have decided to use the variables in a
  non-parametric test).
- Mann–Whitney U test (for use with one categorical variable with two levels, and
  one continuous variable that you have decided to use in a non-parametric test).
- Kruskal–Wallis H test (for use with one categorical variable with three levels or
  more, and one continuous variable that you have decided to use in a non-
  parametric test).
- Friedman test (when you have the same continuous variable administered on
  three occasions, that you have decided is suitable for use in a non-parametric
  test).

In this chapter, aspects of SPSS for Windows will be illustrated using the dataset
'Datafile.sav', so ensure you load up this dataset when using SPSS for Windows.

Exercise 1: Chapter energiser

Try the Minefield Puzzle given in Figure 6.1. This puzzle is designed to put you in the
frame of mind for tackling statistical problems.

Figure 6.1   Minefield Puzzle.

The grid below shows boxes with numbers, and boxes that are blank. Within the blank boxes, 10 of them contain bombs.

The numbers tell you how many bombs are in the area immediately touching it (this can be vertically, horizontally, or diagonally).

Can you identify where the 10 bombs are?

Example
Mark each bomb with a cross, and shade in each empty box as shown in the example

2	2	1	
x	x	1	

	1	2	2	1			
	2			1			
	2		4	2	1		
	1	1	2		1		
			2	2	2		
1	1	2	2		1		
2		2		3	2	2	1
	2	2	1	2		2	

Non-parametric tests

As described in Chapter 5, non-parametric tests are used when the continuous data does not meet the assumptions needed for using a parametric test. As such, non-parametric tests are a range of tests fulfilling two functions of significance testing. The first provides a test for use when a researcher wants to examine the relationship between two categorical variables (chi-square test). The second function is that a number of non-parametric tests provide an alternative to the parametric tests described in Chapter 6, when the researcher decides that the continuous data being used does not fulfil the criteria for use in parametric tests.

Non-parametric test 1: chi-square

The chi-square statistic is performed when you want to examine the relationship between two categorical variables. For example, if we had two variables, Sex of respondent and Type of work performed by the respondent (paid and unpaid work), we would be able to examine whether the two variables are associated. We use the chi-square test to determine whether there is a significant association between the two variables.

The key idea underlying the chi-square test is that we examine two sets of frequencies, observed and expected. Table 6.1 begins to represent the ideas underlying the chi-square, by setting out the findings of a researcher from a study of 100 respondents in which respondents were asked about their sex and whether they were employed in paid or unpaid work. The chi-square forms a matrix made up of cells representing each of the possible combinations of the levels of each categorical variable (i.e. males and paid work, females and paid work, males and unpaid work, and females and unpaid work). As you can see from the table, there are two possible sets of frequencies.

The first set of numbers (in brackets) is the 'expected frequencies'. These are an example of a set of frequencies the researcher would expect to emerge if there was no association between the two variables, with the frequencies spread evenly across the possible combination of levels for each of the categorical variables (males and paid work = 25, females and paid work = 25, males and unpaid work = 25, and females and unpaid work = 25).

The second set of numbers is the 'observed frequencies'. These are an example of a set of frequencies the researcher might have found having gone out and collected the data (males and paid work = 40, females and paid work = 10, males and unpaid work = 10, and females and unpaid work = 40).

The chi-square test uses significance testing to examine whether the observed frequencies (the data collected by the researcher) fit with what would be normally be expected. If the data does not fit (a significant result) through there being an over/under-representation of respondents in one or more of the cells, the researcher would conclude that there is a significant association between the two variables.

Table 6.1   Breakdown of 100 respondents by sex and type of work.

	Paid work	Not in paid work
Males	40 (25)	10 (25)
Females	10 (25)	40 (25)

Take, for example, the data in Table 6.1 for Sex and Type of work. If these variables were not related we would expect the 100 respondents to be distributed fairly equally in each of the cells (i.e. 25 respondents in each). Here, however, we find that males tend to be in paid work, while females tend not to be in paid work. This sort of finding would be an example of a significant finding using chi-square. That is, there is an association between your gender and what type of work you do.

Performing the chi-square test in SPSS for Windows

'The greater religiosity of women must be one of the oldest, and clearest, findings in the psychology of religion' (Beit-Hallahmi and Argyle, 1997, p. 142).

Beit-Hallahmi and Argyle illustrate this conclusion by reporting a number of studies showing that women engage more in religious rituals (Fichter, 1952), and daily prayer (Gallup, 1980) and demonstrate higher levels of religious belief (Gerard, 1985; Yeaman, 1987) than men. In the last chapter, you were introduced to the variable religious type, i.e. whether respondents were extrinsically religious (publicly religious), intrinsically (personally religious) or non-religious. Therefore, we can examine Beit-Hallahmi and Argyle's conclusion with our present data. However, our research examines whether there is a significant association between sex and religious type of respondents.

In this example we are going to examine the relationship between the SEX variable (1 = male, 2 = female) and the RELTYPE variable (religious type: 3 = extrinsic, 2 = intrinsic, 1 = non-religious). Click on the **Analyse** pull-down menu, click on **Descriptive Statistics**, and then click on **Crosstabs**. You will get a Window like Figure 6.2. Within the **variables** box in the Crosstabs window, click on SEX and then on > to move the variable into the **Row(s)** box. Click on RELTYPE and then on > to move the variable into the **Column(s)** box. Then click on the **Statistics...** box.

Click the box next to the chi-square. Click on **Continue**, and then click on the **Cells...** button. Here, click total in the **Percentages**. Again click on **Continue** and then run the CROSSTABS procedure by pressing **OK**. You should now get an output window like that in Table 6.2.

In this output we have all the information we need to interpret whether there is a significant association between sex and religious type. Again, the important rule to remember when interpreting and writing tests is to *Describe* and then *Decide*. That is, describe what is happening within the findings, and then decide whether the result is significant.

Figure 6.2   Crosstabs window.

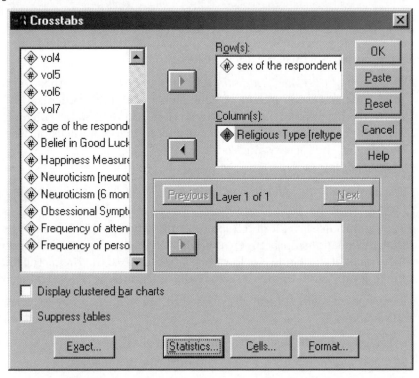

Table 6.2   Chi-square output.

Sex of the respondent* Religious type cross tabulation						
			Religious type			
			Non-religious	Intrinsic	Extrinsic	Total
Sex of the respondent	Male	Count	9	14	6	29
		% of total	15.0%	23.3%	10.0%	48.3%
	Female	Count	9	10	12	31
		% of total	15.0%	16.7%	20.0%	51.7%
Total		Count	18	24	18	60
		% of total	30.0%	40.0%	30.0%	100.0%

Chi-square tests			
	Value	df	Asymp. sig. (2–sided)
Pearson chi-square	2.603[a]	2	0.272
Likelihood ratio	2.642	2	0.267
Linear-by-linear association	0.984	1	0.321
N of valid cases	60		

[a] 0 cells (0%) have expected count less than 5. The minimum expected count is 8.70.

Using Describe and Decide to interpret the chi-square test

From the output, you will need to consider the following:

1. *Breakdown of the values within the cells.* It is important here that you try to provide an overall picture of where most of the respondents, or not many of the respondents, are placed. Here, we deliberately include a percentage breakdown for the overall total to help you focus on where the highest and lowest values occur.
2. *Pearson chi-square.* The statistical test statistic.
3. *The Asymp. Sig. (two-tailed).* The significance level. This is the probability level given to the current findings.
4. The significance level found in point 2, and whether this figure suggests that the relationship between the two variables is significant, or not. Remember, if this figure is below the $p = 0.05$ or $p = 0.01$ criteria, the finding is significant. If this figure is above 0.05, the finding is not significant.

There are six cells which break down each of the frequencies by each level of the variables:

- 9 (15.0%) respondents are male and non-religious;
- 14 (23.3%) respondents are male and intrinsically religious;
- 6 (10.0%) respondents are male and extrinsically religious;
- 9 (15.0%) respondents are female and non-religious;
- 10 (16.7%) respondents are female and intrinsically religious;
- 12 (20.0%) respondents are female and extrinsically religious.

At present, the distribution among the cells seems evenly split. If there is an association between the variables, this would, at best, suggest that a higher percentage of men are intrinsically religious, and less extrinsically religious. However, the way to determine whether there is an association is to use significance testing. The significance level is $p = 0.272$. This is above the criterion of 0.05, therefore we can conclude that there is no significant association between sex and religious type. We have found evidence, therefore, to suggest that, when examining different religious orientations, we find no differences between males and females.

Using Decide and Describe to report the chi-square

The next stage is that you will need to report this statistic. There is a formal way of reporting the chi-square. This comprises two elements. First there is a formal statement of your statistics, and must include:

- The test statistic. Each test has a symbol for its statistic. The chi-square symbol is χ^2. Therefore, in your write-up you must include what χ^2 equals. In the example $\chi^2 = 2.603$.

- The degrees of freedom. This is the number of rows minus 1, times the number of columns minus 1 (here $2 - 1 \times 3 - 1 = 1 \times 2 = 2$). This is placed between the χ^2 and the = sign and placed in brackets. Here, the degrees of freedom are 2 (here $2 - 1 \times 3 - 2 = 1 \times 2 = 2$). Therefore, $\chi^2(2) = 2.603$.

- Third, you must report the probability. This is done in relation to whether your probability value was below 0.05 or 0.01 (significant) or above 0.05 (not significant). Here, you use less than (<) or greater than (>) the criteria level. You state this criteria by stating whether $p < 0.05$ (significant), $p < 0.01$ (significant) or $p > 0.05$ (not significant). So, in the example above, as $p = 0.272$, we would write $p > 0.05$ and place this after the reporting of the χ^2 value. Therefore, $\chi^2(2) = 2.603, p > 0.05$.

This must then be incorporated into the text, to help the reader understand and conceptualise your findings. In writing the text use the *Describe* and *Decide* rule to inform your reader of your finding:

- Remind the reader of the two variables you are examining.
- Describe the relationship between the two variables in terms of the cell counts (and percentages).
- Tell the reader whether the finding is significant or not.

You can use all the information above to write a fairly simple sentence, which conveys your findings succinctly, but effectively. However, you will also need to include a table. Therefore, using the findings above we might report:

Table 6.3 shows a breakdown of the distribution of respondents by each of the cells (with percentages in brackets). The frequency of respondents in each cell was evenly distributed with the greatest number of respondents falling in the male and intrinsic religion cell (14, 23.3%) and the lowest number of respondents falling within the male and non-religious cell. A chi-square was used to determine whether there was a significant association between the two variables. No significant association was found between sex and religious type ($\chi^2(2) = 2.603, p > 0.05$).

Table 6.3 Example of table for reporting the chi-square: breakdown of sample by sex and religious type.

			Religious type	
		Extrinsic	Intrinsic	Non-religious
Sex of respondent	Male	9 (15.0%)	14 (23.3%)	6 (10.0%)
	Female	9 (15.0%)	10 (16.7%)	12 (20.0%)

Writing a significant result for the chi-square

Example: A chi-square was used to determine whether there was a significant association between the two variables. A significant association was found between sex and religious type ($\chi^2(2) = 6.603$, $p < 0.05$).

You will of course find different ways of writing up the chi-square statistic, both in your writing and by other authors, but you will find all the information included above in each write-up.

Further information on the chi-square (χ^2) test

There are two conditions under which it is not advisable to do a chi-square test.

1. When you have a chi-square that comprises 2 rows by 2 columns and any of the frequencies are less than 5.
2. When you have a chi-square that comprises more than 2 rows by 2 columns and any of the frequencies are less than 1 or more than 20 per cent of the frequencies are less than 5.

Figure 6.3   Calculating the chi-square by hand.

Let us work through an example of 100 students. We have a breakdown of their sex and their university degree. As such, we have two categorical variables, sex of respondent and university course. Let us examine whether there is an association between sex and choice of degree.

	Science	Arts	Row totals
Male	32 (A: Step 1)	15 (B: Step 1)	47 (Step 2)
Female	20 (C: Step 1)	33 (D: Step 1)	53 (Step 2)
Column and row total	52 (Step 3)	48 (Step 3)	Overall total = 100 (Step 4)

Step 1: Name each of the cells:
32 = A, 15 = B, 20 = C, 33 = D.

Step 2: Work out the total for each of the rows:
32 + 15 = 47, 20 + 33 = 53

Step 3: Work out the total for each of the columns:
32 + 20 = 52, 15 + 33 = 48

continued

Figure 6.3   (*continued*)

Step 4:   Work out the total for all the cells:
32 + 15 + 20 + 33 = 100

Step 5:   Work out expected frequency for each of the cells, by multiplying its row
total by its column total divided by the total for the sample.
A = 47 × 52/100 = 24.44, B = 47 × 48/100 = 22.56, C = 53 × 52/100 = 27.56,
D = 53 × 48/100 = 25.44

	Observed	Expected (Step 5)	Step 6	Step 7	Step 8
A	32	24.44	7.56	57.1536	2.339
B	15	22.56	−7.56	57.1536	2.533
C	20	27.56	−7.56	57.1536	2.074
D	33	25.44	7.56	57.1536	2.247

Step 6:   Subtract expected frequency from the observed value:
A: 32 − 24.44 = 7.56, B = 15 − 22.56 = −7.56, C = 20 − 27.56 = −7.56,
D = 33 − 25.44 = 7.56

Step 7:   Square each of the values obtained in Step 6: Each is 57.1536

Step 8:   Divide your finding for Step 7 by your finding for each cell in Step 5
(3 dec places):
57.1536/24.44 = 2.339, 57.1536/22.56 = 2.533, 57.1536/27.56 = 2.074,
57.1536/25.44 = 2.247

Step 9:   Add up all your scores for all your cells in Step 8:
2.339 + 2.533 + 2.074 + 2.247 = 9.193. This is your chi-square statistic.

Step 10: Work out your degrees of freedom by multiplying number of rows (−1) by
number of columns (−1) . . . . . . (2 − 1) × (2 − 1) = 1 × 1 = 1.

Step 11: Look at each row and find the value that the chi-square is bigger than.
Here it is bigger than 6.64, so our finding is significant at the 0.001 level.

	Significance levels for 1–tailed test				
	0.05	0.025	0.01	0.005	
	Significance levels for 2–tailed test				
df	0.10	0.05	0.02	0.01	
1	2.71	3.84	5.41	6.64	
2	4.60	5.99	7.82	9.21	
3	6.25	7.82	9.84	11.34	
4	7.78	9.49	11.67	13.28	

Exercise 2: Assessment check for chi-square

Table 6.4 shows the SPSS for Windows Output for a chi-square being used to examine whether a significant association occurs, among 50 adults, between sex and whether respondents categorised themselves as either (1) a happy person or (2) an unhappy person. The table includes a breakdown of frequencies and the chi-square statistic. Using this information, determine whether a significant difference occurs between sex and the description of happiness.

Table 6.4   Assessment check for chi-square.

SEX * HAPPY cross tabulation

Count

		HAPPY		
		Happy	Unhappy	Total
SEX	Male	7	16	23
	Female	18	9	27
Total		25	25	50

Chi-square tests

	Value	df	Asymp. sig. (2-sided)	Exact sig. (2-sided)	Exact sig. (1-sided)
Pearson chi-square	6.522[b]	1	0.011		
Continuity correction[a]	5.153	1	0.023		
Likelihood ratio	6.676	1	0.010		
Fisher's exact test				0.022	0.011
Linear-by-linear association	6.391	1	0.011		
N of valid cases	50				

[a] Computed only for a 2 × 2 table.
[b] 0 cells (0%) have expected count less than 5. The minimum expected count is 11.50.

Non-parametric alternatives to parametric tests

Throughout the rest of this chapter we are going to cover those tests that are non-parametric alternatives to the tests introduced in Chapter 5. These are tests that you use when you are unable to assume that your continuous data demonstrates parametric properties. Therefore, we will introduce you to five other non-parametric tests:

- Spearman correlation – the non-parametric version of the Pearson product–moment correlation coefficient.

- Wilcoxon sign-ranks – the non-parametric version of the related *t*-test.

- Mann–Whitney U – the non-parametric version of the independent-sample *t*-test.

- Kruskal–Wallis H test – the non-parametric version of the analysis of variance – between subjects.

- Friedman test – the non-parametric version of the analysis of variance – within subjects.

These tests differ from parametric tests in that, for continuous data, the numbers within data are usually ranked, from lowest to highest. So, for example, if scores on the test were 3, 7, 17, 33, 45, they would be ranked as 1, 2, 3, 4 and 5. This is done to solve some of the uncertainty surrounding this type of data. Remember, data is used in non-parametric tests because the researcher is not confident the data can be viewed as demonstrating particular numerical properties; for example, the values assigned represent artificial categories (1 = not at all, 2 = sometimes, 3 = often, 4 = always) or scores are grouped together in a skewed distribution. By ranking the data you are assigning some numerical order to the scores.

Non-parametric test 2: Spearman's rho correlation (alternative to the Pearson correlation)

The Spearman correlation is the non-parametric version of the parametric test, the Pearson product–moment correlation. Whereas the Pearson product–moment correlation is used to examine whether a significant relationship occurs between two continuous variables which demonstrate properties suitable for use in parametric tests, the Spearman rho correlation is used to examine whether a significant relationship occurs between two continuous variables which demonstrate properties suitable for use in non-parametric tests. Apart from this, the Spearman rho correlation works in exactly the same way, with an *r* value between −1.00 and +1.00 being generated, and significance testing being used to determine whether a significant relationship occurs between two variables.

Doing the Spearman rho correlation in SPSS for Windows

For this example, we are going to use two variables that measure two aspects of religiosity, (i) frequency of attendance at a place of worship (FREQPOW: 1 = never, 2 = rarely, 3 = monthly, 4 = weekly, 5 = once a week or more) and

(ii) frequency of personal prayer (FREQPP: 1 = never, 2 = rarely, 3 = sometimes, 4 = weekly, 5 = daily). These are both scales that use ordinal data (where numbers are assigned to categories of the variable, but these are artificial numbers); this is one of the reasons why some researchers would use a non-parametric test.

Before we confirm the use of a non-parametric test, let us take a further look at these two variables. Figure 6.4 shows the distribution of scores for Frequency of attendance at a place of worship and Frequency of personal prayer. Both histograms show positively skewed data, with a greater frequency for lower scores. Remember, one of the assumptions for parametric testing is that data demonstrates a normal distribution. Therefore, this finding might be another reason why some researchers might use a non-parametric test.

Figure 6.4   Histogram to show the positive skew of the distribution of scores for (a) the Frequency of attendance at place of worship variable and (b) the Frequency of personal prayer variable.

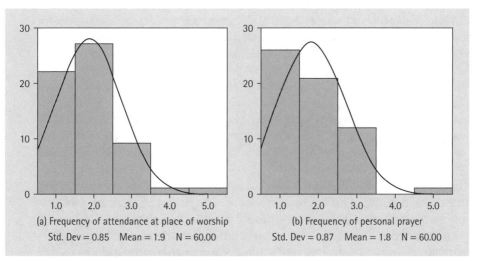

(a) Frequency of attendance at place of worship
Std. Dev = 0.85   Mean = 1.9   N = 60.00

(b) Frequency of personal prayer
Std. Dev = 0.87   Mean = 1.8   N = 60.00

Obtaining the skewed statistic on SPSS for Windows

To obtain the skewed statistic click on the **Analyse** pull-down menu, click on **Descriptive Statistics**, then on **Descriptives**. Transfer your variables into the variable box and click on **Options**. In this window, you will see a number of Descriptive Statistics already ticked: Means, Standard Deviation, Minimum (lowest score obtained in sample), and Maximum (highest score obtained in sample). To get the skewness statistic, click the Skewness statistic here. Press **Continue** and then **OK**.

Table 6.5 Skew statistics for Frequency of attendance at place of worship and Frequency of personal prayer variables.

						Descriptive statistics		
	N	Minimum	Maximum	Mean	Std.		Skewness	
	Statistic	Statistic	Statistic	Statistic	Statistic	Statistic	Std. error	
FREQCA	60	1.00	5.00	1.8667	0.8530	1.111	0.309
FREQPP	60	1.00	5.00	1.8167	0.8732	1.004	0.309
Valid N (listwise)	60						

Finally, as a way to confirm skewness other than from a graph, Table 6.5 shows that the mean score statistic for both variables included in this table is a skew statistic. For the Frequency of attendance at a place of worship the skew statistic is 1.111 and that for the Frequency of personal prayer is 1.004. One criterion used to determine whether data is skewed is whether the skewness statistic is over 1. If the skewness statistic is over 1 then the data is skewed. Both these statistics suggest the data is skewed. Some researchers would suggest that, as the data is skewed, it is not demonstrating properties suitable for parametric testing, and therefore, it would be prudent to use a non-parametric test.

To compute and interpret this correlation coefficient in SPSS for Windows, you generally do as you would for the calculation of the Pearson product–moment correlation. Pull down the **Analyse** menu, click on **Correlate**, then **Bivariate**. Transfer the two variables (FREQPOW; FREQPP) into the Variables box. However, in the **Bivariate Correlations** window (when choosing your variables) choose Spearman rather than the Pearson box in the bottom left hand corner of the window. Press **OK** and you will get the output shown in Table 6.6.

Table 6.6 Spearman's rho output.

			Correlations	
			Frequency of attendance at place of worship	Frequency of personal prayer
Spearman's rho	Frequency of attendance at place of worship	Correlation coefficient	1.000	0.521**
		Sig. (2-tailed)	.	0.000
		N	60	60
	Frequency of personal prayer	Correlation coefficient	0.521**	1.000
		Sig. (2-tailed)	0.000	.
		N	60	60

** Correlation is significant at the 0.01 level (2-tailed).

In this output we have all the information we need to interpret whether there is a relationship between Frequency of attendance at a place of worship and Frequency of personal prayer. The important rule to remember when interpreting and writing tests is to *Describe* and then *Decide*. That is, describe what is happening within the findings, and then decide whether the result is significant.

Using Describe and Decide to interpret the test

From the output, you will need to consider three things:

- *Spearman's rho*. The statistical test. It is important to note whether the statistic is positive or minus. This indicates whether the relationship between the two variables is positive (positive number, though SPSS doesn't print a +) or negative (represented by a – sign).
- *The Sig. (2-tailed)*. The significance level. This is the probability level given to the current findings.
- The significance level found in bullet-point 2 and whether the figure suggests that the relationship between the two variables is significant or not. Remember, if this figure is below the $p = 0.05$ or $p = 0.01$ criteria, the finding is significant. If this figure is above 0.05, the finding is not significant.

Therefore, the correlation between the variables Frequency of personal prayer and Frequency of attendance at a place of worship is $r = 0.521$. This tells us that there is a positive relationship between frequency of personal prayer and frequency of church attendance (if it was negative the statistics would have a minus sign in front of it). The significance level is $p = 0.000$. This is below 0.01, therefore, we can conclude that there is a *significant* positive relationship between frequency of personal prayer and frequency of attendance at a place of worship. This means the more that respondents go to church, the more they practice prayer.

Using Decide and Describe to report the Spearman's rho correlation

The next stage is that you will need to report this statistic. There is a formal way of reporting the Spearman's rho correlation. This comprises two elements. First there is a formal statement of your statistics, and must include:

- The test statistic. Each test has a symbol for its statistic. The Spearman's rho correlation uses the symbol r. Therefore, in your write-up you must include what r equals. In the example above $r = 0.521$.
- The degrees of freedom. This was introduced in Chapter 5, and is traditionally reported (though it is worth noting that it is not always reported). For the Spearman's rho the degrees of freedom equal the size

of your sample minus 2. Here, the minus 2 represents minus 1 for each set of scores, the set of scores for frequency of personal prayer and the set of scores for attendance at a place of worship. This is placed between the r and the = sign and placed in brackets. Here, the degrees of freedom are 58 (size of sample = 60, minus 2 = 58). Therefore $r(58) = 0.521$.

- Third, you must report the probability. This is done in relation to whether your probability value was below 0.05 or 0.01 (significant) or above 0.05 (not significant). Here, you use less than (<) or greater than (>) the criteria level. You state this criterion by stating whether $p < 0.05$ (significant), $p < 0.01$ (significant) or $p > 0.05$ (not significant). In the example above, as $p = 0.000$, we would write $p < 0.01$ and place this after the reporting of the r value. Therefore, in our findings, $r(58) = 0.521, p < 0.05$.

This must then be incorporated into the text, to help the reader understand and conceptualise your findings. In writing the text use the *Describe* and *Decide* rule to inform the reader of your finding:

- Remind the reader of the two variables you are examining.
- Describe whether the relationship between the two variables is positive or negative.
- Tell the reader whether the finding is significant or not.

You can use all the information above to write a fairly simple sentence, which conveys your findings succinctly but effectively. Therefore, using the findings above we might report:

> *A Spearman's rho correlation was used to examine the relationship between Frequency of personal prayer and Frequency of church attendance. A significant positive correlation was found between Frequency of personal prayer and Frequency of attendance at a place of worship (r[58] = 0.521, p < 0.01).*

Writing a non-significant result for Spearman's rho correlation coefficient

Example: A Spearman's rho correlation was used to examine the relationship between scores on the Frequency of personal prayer and Frequency of attendance at a place of worship. No significant correlation was found between frequencies of personal prayer and attendance at a place of worship ($r[58] = 0.043$, $p > 0.05$).

You will of course find different ways of writing up the Spearman rho correlation, both in your writing and by other authors, but you will find all the information included above.

Figure 6.5 Calculating the Spearman's rho correlation by hand.

The aim of the present calculation is to examine whether we can replicate the findings of a significant positive association between Frequency of attendance at a place of worship (FREQPOW: 1 = never, 2 = rarely, 3 = monthly, 4 = weekly, 5 = once a week or more) and Frequency of personal prayer (FREQPP: 1 = never, 2 = rarely, 3 = sometimes, 4 = weekly, 5 = daily) among five respondents.

Respondent	FFPP	FFPOW	Rank PP Step 1	Rank POW Step 1	Difference Step 2	Difference² Step 3
1	2	1	2.5	1	+1.5	2.25
2	1	3	1	4.5	−3.5	12.25
3	3	2	4.5	2.5	+2.0	4
4	4	4	6	5.5	+0.5	0.25
5	5	4	7	5.5	+1.5	2.25
6	3	2	4.5	2.5	+2	4
7	2	3	2.5	4.5	−2	4

Step 1: Rank the scores for Frequency of personal prayer and Frequency of place of worship in separate columns. If two people have exactly the same ranks, e.g. Respondents 1 and 7 have same score for FFPP then add ranks 2 + 3 (these would be ranked 2nd and 3rd) and divide by number of respondents: 2 --- 2 + 3, 9/2 = 2.5.

Step 2: Subtract Rank POW from the Ranks PP:
Respondent 1: 2.5 − 1 = 1.5
Respondent 2: 1 − 4.5 = −3.5
Respondent 3: 4.5 − 2.5 = 2
Respondent 4: 6 − 5.5 = +0.5
Respondent 5: 7 − 5.5 = +1.5
Respondent 6: 4.5 − 2.5 = 2
Respondent 7: 2.5 − 4.5 = −2

Step 3: Square each of the findings for Step 2:
Respondent 1: 1.5 × 1.5 = 2.25
Respondent 2: −3.5 × −3.5 = 12.25
Respondent 3: 2 × 2 = 4
Respondent 4: +0.5 × +0.5 = 0.25
Respondent 5: +1.5 × +1.5 = 2.25
Respondent 6: +2 × +2 = 4
Respondent 7: −2 × −2 = 4

Step 4: Add together the results of Step 3:
2.25 + 12.25 + 4 + 0.25 + 2.25 + 4 + 4 = 29.5

Step 5: Multiply the findings of Steps 4 and 6:
29.5 × 6 = 177

Step 6: Square the number of people in the sample and then minus 1:
(7 × 7) − 1 = 49 − 1 = 48

Step 7: Multiply the number of people in the sample by your finding for Step 6:
7 × 48 = 336.

continued

Figure 6.5 (*continued*)

Step 8: Divide your finding for Step 5 by your finding for Step 7:
177/336 = 0.5268

Step 9: Subtract your finding for Step 8 from 1:
1 – 0.5268 = 0.9583; this is your Spearman rho statistic.
Spearman rho = 0.4732

Step 10: Work out your sample size: N = 7
7 – 2 = 5. This is your degrees of freedom.

Step 11: Look at the table and determine the number that accompanies two-tailed test (we have made no particular statement regarding the direction of the relation between the two variables) and for a criteria level of 0.05. This number is 0.714. Spearman rho (r = 0.4732) is not larger than 0.714. Therefore, our finding is not significant.

	Significance levels for 1–tailed test				
	0.05	0.025	0.01	0.005	
	Significance levels for 2–tailed test				
df	0.10	0.05	0.02	0.01	
5	0.900	1.000	1.000		
6	0.829	0.886	0.943	1.000	
7	0.714	0.786	0.893	0.929	
8	0.643	0.738	0.833	0.881	

Non-parametric test 3: Wilcoxon sign-ranks test (alternative to the related *t*-test)

This is the non-parametric alternative to the related t-test. This test is used when you have measured the same continuous variable on two occasions among the same respondents, and you do not wish to treat your continuous data as parametric. While the related *t*-test was based on examining whether significant differences occur between mean scores (with standard deviations), the Wilcoxon sign-ranks test, because data is ranked, is based on examining significant differences between average ranks (mean ranks).

Performing the Wilcoxon sign-ranks test in SPSS for Windows

Let us use the following example. Using the variable Frequency of attendance at a place of worship (FREQPOW: 1 = never, 2 = rarely, 3 = monthly, 4 = weekly, 5 = once a week or more) we administered measures of this variable twice, the first time with the original questionnaire (FREQPOW) and a second time, six months later (FREQPOW2). Using these two variables we can see whether respondents significantly differed in their religious behaviour over six months.

To perform the Wilcoxon sign-ranks test go to the **Analyse** pull-down menu and click on **Nonparametric** tests. Next click onto **2 related samples**. Highlight the two variable names in the left hand box, and they will appear in the **Current Selections** box below. Then move them into the **Test Pair(s) List:** box by clicking on the > (see Figure 6.6). Make sure the Wilcoxon box is ticked and then click on **OK**. You will then get an output like Table 6.7.

In this output we have all the information we need to interpret whether a significant difference occurs between scores for the two administrations of the attendance at the place of worship variables. Again, remember to *Describe* and then *Decide*.

Using Describe and Decide to interpret the Wilcoxon sign-ranks test

From the output, you will need to consider three things:

• *Mean ranks*. These are the basis of our description. Here, lower ranks mean lower scores (remember 2, 5, 7, 10, 13 would be ranked 1, 2, 3, 4, 5). We note both the mean ranks, and then, which mean rank is higher. In the Wilcoxon sign-ranks test output in SPSS for Windows this is expressed in a certain way. If you look at Table 6.7 you will see that in the Ranks table, in the N column there are two letters, a and b. These correspond to some statements under the table: (a) Frequency of attendance at place of worship – 6 months later < Frequency of attendance at place of worship;

Figure 6.6   Two-Related-Samples Tests window.

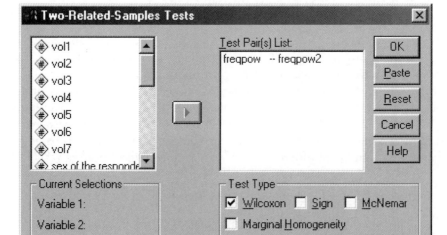

Table 6.7   Wilcoxon sign-ranks test output.

Ranks

		N	Mean rank	Sum of ranks
Frequency of attendance at place of worship – 6 months later – Frequency of attendance at place of worship	Negative ranks	9[a]	7.89	71.00
	Positive ranks	7[b]	9.29	65.00
	Ties	44[c]		
	Total	60		

[a] Frequency of attendance at place of worship – 6 months later < Frequency of attendance at place of worship.
[b] Frequency of attendance at place of worship – 6 months later > Frequency of attendance at place of worship.
[c] Frequency of attendance at place of worship = Frequency of attendance at place of worship – 6 months later.

Test statistics[b]

	Frequency of attendance at place of worship – 6 months later – Frequency of attendance at place of worship
Z	−0.166[a]
Asymp. Sig. (2-tailed)	0.868

[a] Based on positive ranks.
[b] Wilcoxon sign-ranks test.

(b) Frequency of attendance at place of worship – 6 months later > Frequency of attendance at place of worship. To simplify a complicated process (which is not covered in SPSS for Windows), determine your mean ranks for each variable by assigning a mean rank to whichever variable is greater than in the statements below the table. So, the mean rank for the variable Frequency of attendance at place of worship is the row with (a) (mean rank = 7.89); the mean rank for the variable Frequency of attendance at place of worship – 6 months later is in the row with (b) (mean rank = 9.29).

- *The z value.* The test statistic.
- *The Asymp. Sig. (2-tailed).* The significance level. This is the probability level given to the current findings. The significance level indicates whether there are significant differences noted between the mean scores. Remember, if this figure is below the $p = 0.05$ or $p = 0.01$ criteria, the finding is significant. If this figure is above 0.05, the finding is not significant.

Therefore, the average mean rank for Frequency of attendance at a place of worship on the first administration is 7.89 and the average mean rank for

Frequency of attendance at a place of worship on the second administration is 9.29. Here, we note that scores for the second administration are higher than those for the first administration. The z value for the statistic is –0.166. This tells us very little at this stage. However, the significance level is $p = 0.868$. This is greater than 0.05 and so we conclude that there is not a significant difference for mean scores between the two administrations of Frequency of attendance at a place of worship. This suggests, from our sample, that respondents do not significantly differ in their religious behaviour over a six-month period.

Using Decide and Describe to report the Wilcoxon sign-ranks test

The next stage is that you will need to report these statistics. There is a formal way of reporting the Wilcoxon sign-ranks test. This comprises two elements. First there is a formal statement of your statistics, and must include:

- The test statistic. Each test has a symbol for its statistic. The Wilcoxon sign-ranks test has the symbol z. Therefore, in your write-up you must include what z equals. In the example $z = –0.166$.

- You must report the probability. Again, this is done in relation to whether your probability value was below 0.05 or 0.01 (significant) or above 0.05 (not significant). Here, you use less than (<) or greater than (>) the criteria level. You state this criteria by stating whether $p < 0.05$ (significant), $p < 0.01$ (significant) or $p > 0.05$ (not significant). So, in the example above, as $p = 0.868$, we would write $p > 0.06$ and place this after the reporting of the z value. Therefore, with our findings, $z = –0.166$, $p > 0.05$.

This must then be incorporated into the text, to help the reader understand and conceptualise your findings. In writing the text, use the *Describe* and *Decide* rule to inform your reader of your finding:

- Remind the reader of the two variables you are examining.
- Describe which mean score is highest.
- Tell the reader whether the finding is significant or not.

You can use all the information above to write a fairly simple sentence, which conveys your findings succinctly, but effectively. Therefore, using the findings above we might report:

A Wilcoxon sign-ranks test was used to examine significant differences between mean ranks on the Frequency of attendance at a place of worship over a six month period. No significant difference ($z = –0.166$, $p > 0.05$) was found for mean scores for Frequency of attendance at a place of worship for the first administration (mean rank = 7.89) and the second administration (mean rank = 9.29).

Writing a significant result for a Wilcoxon pairs test

Example: A Wilcoxon paired test was used to examine significant differences between mean ranks on the Frequency of attendance at a place of worship over a six month period. Mean rank scores for the first administration of Frequency of attendance at a place of worship (mean rank = 10.01) were found to be significantly higher ($z = 5.30$, $p < 0.05$) than scores for the second administration of the measure of Frequency of attendance at a place of worship (mean rank = 8.24).

Again, you will come across different ways of writing tests, but you will find all the information included above in any write-up.

Figure 6.7   Calculating the Wilcoxon pairs test by hand.

The sample of seven respondents were asked to complete the same measure twice on two occasions, six months apart: How often do you feel anxious? 1 = not at all, 2 = very rarely, 3 = sometimes, 4 = monthly, 5 = weekly, 6 = daily, 7 = more than once every day.

Respondent	Anxious time 1	Anxious time 2	Difference	Rank order of differences
1	1	2	−1 (Step 1)	1 (Step 2)
2	4	2	+2 (Step 1)	2.5 (Step 2)
3	4	1	+3 (Step 1)	4 (Step 2)
4	5	5	0 (Step 1)	
5	7	3	+4 (Step 1)	5 (Step 2)
6	2	4	−2 (Step 1)	2.5
7	1	7	−6 (Step 1)	6

Step 1: Work out the differences between each pair of scores:
1 − 2 = −1, 2 − 4 = −2, 4 − 1 = +3, 5 − 5 = 0,
7 − 3 = 4, 2 − 4 = −2, 1 − 7= −6.

Step 2: Rank the order of differences from lowest value to highest value, ignoring 0 (like respondent 4) and ignoring + and − signs. With ranks that are equal, add the next ranks together and divide by the number of respondents who have this rank. For example, respondents 2 and 6 have the second highest rank, so add 2 (the next rank) + 3 (the next rank after that) and divide by the number of respondents. 2 + 3/2 = 5/2 = 2.5, so each person gets a rank of 2.5

Step 3: Add the ranks of the + values and − values:
+ values, 2.5 + 4 + 5 = 11.5 . . . . − values, 1 + 2.5 + 6 = 9.5

Step 4: The smaller set of scores is your Wilcoxon pair value (z)
Here z = 9.5.

continued

Figure 6.7 (*continued*)

To work out whether the result is significant, find the row that is equal to the sample size. Here, we have made no prediction about direction of the differences between the scores on both the variables so we use the two-tailed test. Here, sample size = 7 (N = 7). Find the number in this row that z is bigger than. If z is bigger than all the numbers then z is not significant.

	Significance levels for 1–tailed test			
	0.05	0.025	0.01	0.0001
	Significance levels for 2–tailed test			
Sample size	0.10	0.05	0.02	0.0002
5	$T < 0$			
6	2	0		
7	3	2	0	
8	5	3	2	1
9	8	5	3	3

Table 6.8 Assessment check for Wilcoxon sign-ranks test.

		N	Mean rank	Sum of ranks
	Ranks			
Kindness after the film	Negative ranks	36[a]	24.17	870.00
Kindness before the film	Positive ranks	8[b]	15.00	120.00
	Ties	6[c]		
	Total	50		

[a] Kindness after the film < Kindness before the film.
[b] Kindness after the film > Kindness before the film.
[c] Kindness before the film = Kindness after the film.

Test statistics[b]	
	Kindness after the film − Kindness before the film
z	−4.544[a]
Asymp. sig. (2-tailed)	0.000

[a] Based on positive ranks.
[b] Wilcoxon sign-ranks test.

Exercise 3: Assessment check for Wilcoxon sign-ranks test

Table 6.8 shows the SPSS for Windows output for a Wilcoxon sign-ranks test being used to examine whether a significant difference occurs for mean ranks for two administrations of kindness scale, before and after watching a programme about people working for charities. The kindness scale is a one-item measure in which respondents are asked to respond to the statement 'How kind are you?'. Responses are scored, 1 = not at all, 2 = a little, 3 = somewhat, 4 = very much so. The output includes a breakdown of mean ranks and the Wilcoxon sign-ranks test statistic. Using this information determine whether a significant difference occurs for self-rating of kindness before and after the watching of a film.

Non-parametric test 4: Mann–Whitney U test (the non-parametric alternative to the independent-samples *t*-test)

This is the non-parametric alternative to the independent-samples *t*-test. The independent-samples *t*-test is used to examine for significant differences in means scores of two levels of an independent (categorical) variable (e.g. male and female scores in Belief in Good Luck). The Mann–Whitney U test is used to examine for significant differences on non-parametric continuous data between two levels of a categorical variable.

Performing the Mann–Whitney U test in SPSS for Windows

To illustrate the Mann–Whitney U test we are going to look at sex differences in religion using a general measure of religiosity. In the chi-square example, we found no relationship between gender and religious type. In this example, we are going to use a more general measure of religiosity. Respondents were asked 'How religious are you?' Responses were scored on a five-point scale: 1 = not a lot, 2 = a little, 3 = slightly, 4 = quite a lot, 5 = very much so. Therefore, we are interested in finding out whether a significant difference occurs between men and women on religiosity.

To perform the Mann–Whitney U test click on the **Analyse** pull-down menu and click on **Nonparametric** tests. Next click on **2 independent samples.** Highlight the continuous variable (RELIGRU) in the left hand box and move it to the **Test Variable(s)** box by clicking on the >. Similarly, highlight the categorical variable (SEX) and transfer it to the **Grouping Variable.** Define the values of the grouping (independent) variable by clicking on **Define Groups**, and clicking 1 (Male) and 2 (Female) into the boxes. Make sure the Mann–Whitney U test box is ticked and click on **OK** to get your output (see Table 6.9).

Table 6.9   Mann–Whitney U test output.

	Sex of the respondent	N	Mean rank	Sum of ranks
		Ranks		
How religious are you? Male		29	23.31	676.00
Female		31	37.23	1154.00
Total		60		

	How religious are you?
	Test statistics[a]
Mann–Whitney U	241.000
Wilcoxon W	676.000
z	−3.242
Asymp. sig. (2-tailed)	0.001

[a] Grouping variable: sex of the respondent.

In this output we have all the information we need to interpret whether a significant difference occurs between scores of religiosity between males and females. Again, remember to *Describe* and then *Decide*.

Using Describe and Decide to interpret the Mann–Whitney U test

From the output, you will need to consider three things:

- *Mean ranks*. These are the basis of our description. Here, lower ranks are given to lower scores: 2, 5, 7, 10, 13 would be ranked 1, 2, 3, 4, 5. Therefore, we note both the mean ranks, and then, which mean rank is higher.
- *The U value*. The test statistic.
- *The Asymp. Sig. (2-tailed)*. The significance level. This is the probability level given to the current findings. The significance level informs the researcher whether there are significant differences noted between the means scores. Remember, if this figure is below the $p = 0.05$ or $p = 0.01$ criteria, the finding is significant. If this figure is above 0.05, the finding is not significant.

Therefore, the average mean rank for males for self-description of religiosity is 23.31 and the average mean rank for females for self-description of religiosity is 37.23. Here, we note that scores for females are higher than for males. The U value for the statistic is 241.00. This tells us very little at this stage. However, the significance level is $p = 0.001$. This is smaller than 0.01 and so we conclude that there is a significant difference between males and females in self-description of religiosity. The mean ranks suggest that females score significantly higher than males in describing their religiosity.

Using Decide and Describe to report the Mann–Whitney U test

The next stage is that you will need to report these statistics. There is a formal way of reporting the Mann–Whitney U test. This comprises two elements. First there is a formal statement of your statistics, which must include:

- The test statistic. Each test has a symbol for its statistic. The Mann–Whitney U test has the symbol U. Therefore, in your write-up you must include what U equals. In the example U = 241.00.

- You must report the probability. Again, this is done in relation to whether your probability value was below 0.05 or 0.01 (significant) or above 0.05 (not significant). Here, you use less than (<) or greater than (>) the criteria level. You state this criterion by stating whether $p < 0.05$ (significant), $p < 0.01$ (significant) or $p > 0.05$ (not significant). In the example above, as $p = 0.001$, we would write $p < 0.01$ and place this after the reporting of the U value. Therefore, with our findings, U = 241.00, $p < 0.01$.

This must then be incorporated into the text, to help the reader understand and conceptualise your findings. In writing the text use the *Describe* and *Decide* rule to inform your reader of your finding:

- Remind the reader of the two variables you are examining.
- Describe which mean score is highest.
- Tell the reader whether the finding is significant or not.

You can use all the information above to write a fairly simple sentence, which conveys your findings succinctly, but effectively. Therefore, using the findings above we might report:

A Mann–Whitney U test was used to examine significant differences between males and females in how they describe their religiosity. Females (mean rank = 23.31) were found to score significantly higher (U = 241.00, $p < 0.01$) than males (mean rank = 37.23) in terms of how religious they described themselves to be.

Writing a non-significant result for a Mann–Whitney U test

Example: A Mann–Whitney U test was used to examine significant differences between males and females in how they describe their religiosity. No significant difference (U = 1.00, $p > 0.05$) was found between females (mean rank = 23.31) and males (mean rank = 23.23) in terms of how religious they described themselves to be.

Again, you will come across different ways of writing tests, but you will find all the information included above in any write-up.

Figure 6.8   Calculating the Mann–Whitney U test by hand.

We know about the sex difference of personal prayer and church attendance. However, how about sex difference of general religiosity? The sample of 10 respondents (5 males and 5 females) was asked to complete the following five point measure. How religious are you?: 1 = not at all, 2 = slightly, 3 = a little, 4 = quite a bit, 5 = a lot.

Males	Score	Overall rank	Females	Score	Overall rank
1	2	3 (Step 1)	1	4	10.5 (Step 1)
2	3	6.5 (Step 1)	2	2	3 (Step 1)
3	2	3 (Step 1)	3	4	10.5 (Step 1)
4	3	6.5 (Step 1)	4	5	13.5 (Step 1)
5	1	1 (Step 1)	5	5	13.5 (Step 1)
6	3	6.5 (Step 1)	4	4	10.5 (Step 1)
7	4	10.5 (Step 1)	5	3	6.5 (Step 1)

Step 1: Rank all the scores in order, assigning 1 to the lowest. With ranks that are equal, add the next ranks together and divide by the number of respondents who have this rank. For example, male respondents 2 and 4 have the second highest rank, so does female respondent 2. So add ranks 2, 3, 4 and divide by the number of respondents: 3 --- 2 + 3 + 4, 9/3 = 3

Step 2: Add up all the ranks for men:
3 + 6.5 + 3 + 6.5 + 1 + 6.5 + 10.5 = 30.5

Step 3: Add up all the ranks for women:
10.5 + 3 + 10.5 + 13.5 + 13.5 + 10.5 + 6.5 = 68

Step 4: Multiply the number of men by the number of women:
7 × 7= 49

Step 5: Multiply the number of men by the number of men +1 and then divide this total by 2:
7 × (7 + 1)/2 = 7 × 8/2 = 56/2 = 28

Step 6: Multiply the number of women by the number of men +1 and then divide this total by 2:
7 × (7 + 1)/2 = 7 × 8/2 = 56/2 = 28

Step 7: Add your finding for Step 4 to your finding for Step 5 and minus your finding for Step 2:
49 + 28 – 30.5 = 77 – 22 = 55

Step 8: Add your finding for Step 4 to your finding for Step 6 and minus your finding for Step 3:
49 + 28 – 68 = 77 – 68 = 9

Step 9: Select the smallest from Steps 7 and 8. This is your U value:
Step 8 is smallest. *U = 9*

To work out whether the result is significant, find the row and column that are equal to the sample size.

continued

Figure 6.8   (*continued*)

First value in each box is 1-tailed test 0.005, 2-tailed test at 0.01
Second value in each box is 1-tailed 0.01, 2-tailed test at 0.02
Third value in each box is 1-tailed 0.025, 2-tailed test at 0.05
Fourth value in each box is 1-tailed 0.05, 2-tailed test at 0.10

			Males	Sample 1		
		4	5	6	7	8
	4	-, -, 0, 1	-, 0, 1, 2	0, 1, 2, 3	0, 1, 3, 4	1, 2, 4, 5
Females	5	-, 0, 1, 2	0, 1, 2, 4	1, 2, 3, 5	1, 3, 5, 6	2, 4, 6, 8
Sample 2	6	0, 1, 2, 3	1, 2, 3, 5	2, 3, 5, 7	3, 4, 6, 8	4, 6, 8, 10
	7	0, 1, 3, 4	1, 3, 5, 6	3, 4, 6, 8	4, 6, 8, 11	6, 7, 10, 13
	8	1, 2, 4, 5	2, 4, 6, 8	4, 6, 8, 10	6, 7, 10, 13	7, 9, 13, 15

Find the box where the number of males = number of females (n = 7 and n = 7) box is highlighted. As we are arguing that women are more religious than men we choose 1-tailed. U = 9, and is lower than the 4th value, so U is significant at the 0.05 level.

Non-parametric test 5: The Kruskal–Wallis H test (the non-parametric alternative to the analysis of variance – between subjects)

The Kruskal–Wallis H test is used when you want to compare scores on a continuous variable (that you have determined is suitable for using with a non-parametric test) by the levels of a categorical variable with three levels or more (remember, a Mann–Whitney U test was used when the categorical variable had two levels). As such, the Kruskal–Wallis H test is the non-parametric version of the analysis of variance – between subjects test.

From this point on the procedure and interpretation of the test is similar to other non-parametric statistical tests, and we can illustrate this using two of the variables we have used in previous examples in this chapter. The first variable is RELTYPE, identification of religious type of the person. This is a categorical variable that indicates whether people are either extrinsically religious, intrinsically religious, or not religious (we have done the hard work and used a religiosity scale (Gorsuch and Venable, 1983) to determine these categories). In the present sample, an extrinsically religious person is scored as '3', an intrinsically religious person is scored as '2', and a non-religious person is scored as '1'. With this question, respondents were asked 'How religious are you?' Responses were scored on a five-point scale: 1 = not a lot, 2 = a little, 3 = slightly, 4 = quite a lot, 5 = very much so. We are interested in finding out whether a significant difference occurs between the three religious types of religiosity.

Performing the Kruskal–Wallis H test on SPSS for Windows

Pull down the **Analyse** menu and click on **Non-parametric**, and then **K Independent samples.** You should now get a screen that looks like Figure 6.9. Move the dependent variable 'How religious are you' (RELIGRU) into the **Test Variable List:** box. Move the variable Religious type (RELTYPE) into the **Grouping Variable:** box. Click on the **Define Range...** button and you will get the **Several Independent Samples: Define Range** window (see Figure 6.10). This tells the computer how many levels of the categorical variable there are. There are three levels for Religious type (extrinsic, intrinsic, non-religious). Click 1 into Minimum, and 3 into Maximum. Click on **Continue** and then on **OK.** You will then get an output screen as in Table 6.10.

In this output we have all the information we need to interpret whether a significant difference occurs between religious types for their scores on the way respondents describe their religiosity.

Figure 6.9   Tests for Several Independent Samples window.

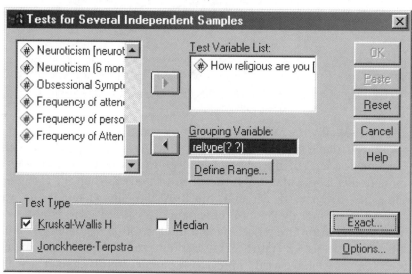

Figure 6.10   Several Independent Samples: Define Range window.

Table 6.10   Kruskal–Wallis H test output.

Ranks			
	Religious type	N	Mean rank
How religious are you?	Non-religious	18	21.61
	Intrinsic	24	33.94
	Extrinsic	18	34.81
	Total	60	

Test statistics[a,b]	
	How religious are you?
Chi-square	7.389
df	2
Asymp. sig.	0.025

[a] Kruskal–Wallis test.
[b] Grouping variable: religious type.

Using Describe and Decide to interpret the Kruskal–Wallis H test

From the first two boxes in this output, you will need to consider three things:

- *Mean ranks* (Table 6.10, Ranks). These are the basis of our description. Here, we note both the mean rank scores (with the standard deviation) by each of the groups of religious persons and note which mean rank score is higher.
- *The chi-square value* (Table 6.10, Test statistics). A chi-square statistic is used for the Kruskal–Wallis H test.
- *The Asymp. Sig.* (Table 6.10, Test statistics). The significance level. This is the probability level given to the current findings. The significance level tells the researcher whether the difference noted between the means is significantly different. Remember, if this figure is below the $p = 0.05$ or $p = 0.01$ criteria, the finding is significant. If this figure is above 0.05, the finding is not significant.

The mean rank for respondents who are non-religious is 21.61, the mean rank for respondents who display intrinsic religiosity is 33.94, and the mean rank for respondents who display extrinsic religiosity is 34.81. Here, we note that extrinsic and intrinsic respondents score higher than non-religious respondents, and that intrinsic and extrinsic mean ranks are not as different. The chi-square statistic is 7.389. This tells us very little at this stage. However, the significance level is $p = 0.025$. This is smaller than 0.05, therefore, we conclude that there is a significant difference for mean ranks in the way respondents describe their religiosity by religious type.

Using post-hoc comparisons with the Kruskal–Wallis H test

Similar to both the analysis of variance – between and within subjects tests, when a significant result is found for the statistical test it is necessary to break down the differences between pairs of mean rank (here, religiosity description comparing extrinsic and intrinsic (1st pair), intrinsic and non-religious (2nd pair) and extrinsic and non-religious (3rd pair). However, unlike the analysis of variance statistical tests, there is no option in the test screen for analysing these differences. Instead, researchers use the Mann–Whitney U test to compare these pairs of differences (remember the Mann–Whitney U test, the one that appears before this one in the chapter, and is used to compare two groups of respondents for their scores on a continuous variable that is suitable for use in a non-parametric test). It is important to note that you would only carry out these series of Mann–Whitney U tests if you found a significant result for the Kruskal–Wallis H test.

Table 6.11 shows a series of Mann–Whitney U tests comparing mean ranks by religious group. These tests show us that a significant difference occurs for self-description of religiosity, between non-religious and intrinsic respondents ($U = 121.50$, $p < 0.05$), between non-religious and extrinsic respondents ($U = 96.50$, $p < 0.05$), but not between intrinsic and extrinsic respondents ($U = 204.00$, $p > 0.05$). These findings suggest that both extrinsic and intrinsic respondents describe themselves as more religious than non-religious respondents, but not more so than each other.

Table 6.11　A series of post-hoc comparisons using the Mann–Whitney U test for religiosity by religious type.

(a) Religiosity by non-religious and intrinsic respondents

Ranks

	Religious type	N	Mean rank	Sum of ranks
How religious are you?	Non-religious	18	16.25	292.50
	Intrinsic	24	25.44	610.50
	Total	42		

Test statistics[a]

	How religious are you?
Mann–Whitney U	121.500
Wilcoxon W	292.500
z	−2.558
Asymp. sig. (2-tailed)	0.011

[a] Grouping variable: religious type.

continued

Table 6.11   (*continued*)

(b) Religiosity by non-religious and extrinsic respondents

Ranks

	Religious type	N	Mean rank	Sum of ranks
How religious are you?	Non-religious	18	14.86	267.50
	Extrinsic	18	22.14	398.50
	Total	36		

Test statistics[b]

	How religious are you?
Mann–Whitney U	96.500
Wilcoxon W	267.500
z	−2.232
Asymp. sig. (2-tailed)	0.026
Exact sig. [2*(1-tailed Sig.)]	0.037[a]

[a] Not corrected for ties.
[b] Grouping variable: religious type.

(c) Religiosity by intrinsic and extrinsic respondents

Ranks

	Religious type	N	Mean rank	Sum of ranks
How religious are you?	Intrinsic	24	21.00	504.00
	Extrinsic	18	22.17	399.00
	Total	42		

Test statistics[a]

	How religious are you?
Mann–Whitney U	204.000
Wilcoxon W	504.000
z	−0.316
Asymp. sig. (2-tailed)	0.752

[a] Grouping variable: religious type.

Using Decide and Describe to report the Kruskal–Wallis H test

The next stage is that you will need to report these statistics. There is a formal way of reporting the Kruskal–Wallis H test. This comprises two elements. First, there is a formal statement of your Kruskal–Wallis H statistic, and must include:

- The test statistic. Each test has a symbol for its statistic. The Kruskal–Wallis H test uses the statistic, χ^2. Therefore, in your write-up you must include what χ^2 equals. In the example above $\chi^2 = 7.3896$.

- The degrees of freedom. Here, the degrees of freedom is 2, therefore $\chi^2(2) = 7.3896$.

- You must report the probability of the Kruskal–Wallis H test. Again, this is done in relation to whether your probability value was below 0.05 or 0.01 (significant) or above 0.05 (not significant). Here, you use less than ($<$) or greater than ($>$) the criteria level. You state this criteria by stating whether $p < 0.05$ (significant), $p < 0.01$ (significant) or $p > 0.05$ (not significant). So, in the example above, as $p = 0.025$, we would write $p < 0.01$ and place this after the reporting of the χ^2 value. Therefore, with our findings, $\chi^2(2) = 7.3896$, $p < 0.05$.

This must then be incorporated into the text, to help the reader understand and conceptualise your findings. In writing the text use the *Describe* and *Decide* rule to inform your reader of your findings:

- Remind the reader of the two variables you are examining.
- Describe which mean score is highest.
- Tell the reader whether the Kruskal–Wallis H test is significant or not.

If the Kruskal–Wallis H test is significant, you must inform the reader where the differences lie between groups. When reporting the series of Mann–Whitney U test statistics you state the test statistic and whether p was smaller than (significant) or greater than (not significant) 0.05. This must then be incorporated into the text, to help the reader understand and conceptualise your findings.

You can use all the information above to write a fairly simple paragraph, which conveys your findings succinctly, but effectively. Therefore, using the findings above you might report:

> *Religious scores were calculated, with mean ranks for the extrinsic group (mean rank = 34.81) and the intrinsic group (mean rank = 33.94), being much higher than the non-religious group (mean rank = 21.61). A Kruskal–Wallis H test indicated a significant difference between the groups on the measure of religious types, $\chi^2(2) = 7.3896$, $p < 0.05$. A series of Mann–Whitney U tests were carried out to provide post-hoc comparisons of the mean ranks. These tests show us*

that a significant difference occurs for self-description of religiosity, between non-religious and intrinsic respondents (U = 121.50, p < 0.05), between non-religious and extrinsic respondents (U = 96.50, p < 0.05), but not between intrinsic and extrinsic respondents (U = 204.00, p > 0.05). These findings suggest that both extrinsic and intrinsic respondents describe themselves as more religious than non-religious respondents, but not more so than each other.

Writing a non-significant result for the Kruskal–Wallis H test

Example: Religious scores were calculated, with mean ranks for the extrinsic group (mean rank = 30.81) and the intrinsic group (mean rank = 30.94), being much higher than the non-religious group (mean rank = 31.61). No significant difference between the groups occurred for scores of how respondents described their religiosity, $\chi^2(2) = 7.3896$, $p < 0.05$.

Again, you will come across different ways of writing tests, but you will find all the information included above in any write-up.

The Jonckheere terpstra test

There is one more piece of information needed. While the Kruskal–Wallis H test is an often used test to examine differences among groups for scores on a continuous variable (that has been considered suitable for using in a non-parametric test) it is essentially a two-tailed test. That is, the researcher has made no specific prediction about the direction of the mean scores (e.g. that one group will score significantly higher than the other groups on the continuous variable). However, it may be that the researcher has made a prediction about the direction of mean scores, essentially meaning the researcher has made a one-tailed prediction. If this is the case the researcher should use the Jonckheere terpstra test, which is a similar test to the Kruskal–Wallis H test, but is a one-tailed test.

To perform the Jonckheere terpstra test on SPSS for Windows, you do exactly the same as you would for the Kruskal–Wallis H test, apart from one thing. When in the Tests for Several Independent Samples window (see Figure 6.9) click the Jonckheere-Terpstra box rather than the Kruskal–Wallis H test. Aside from that, the analysis, interpretation, and write-up (aside from reporting you used the Jonckheere terpstra test rather than the Kruskal–Wallis H test) is the same.

> **Exercise 4:** Self-assessment check for the Kruskal–Wallis H test
>
> Table 6.12 shows the SPSS for Windows output for a Kruskal–Wallis H test being used to examine whether a significant difference occurs between three religious groups (1 = non-religious, 2 = intrinsic, 3 = extrinsic) for scores on the kindness scale. The kindness scale is a one-item measure in which respondents are asked to respond to the statement 'How kind are you?'. Responses are scored, 1 = not at all, 2 = a little, 3 = somewhat, 4 = very much so. The researcher has predicted that non-religious persons will be less kind and therefore has used the Kruskal–Wallis H test. The output includes a breakdown of mean ranks and the Kruskal–Wallis H test statistic. Using this information, determine whether a significant difference occurs between type of religious person and their self-rated level of kindness.

Table 6.12 Self-assessment check for the Kruskal–Wallis H test.

Ranks			
	RELTYPE	N	Mean rank
Kindness scale	Non-religious	12	23.08
	Intrinsic	25	24.54
	Extrinsic	13	29.58
	Total	50	

Test statistics[a,b]	
	Kindness scale
Chi-square	1.629
df	2
Asymp. sig.	0.443

[a] Kruskal–Wallis test.
[b] Grouping variable: RELTYPE.

Non-parametric test 6: Friedman test

The simplest way to view the Friedman test is that it is used when you want to compare scores on the same continuous variable, that you view as being suitable for use in a non-parametric test, that have been administered on three occasions or more. The Friedman test is the non-parametric version of the analysis of variance – within subjects, and works in a similar way by comparing the mean ranks of the same variable over three occasions or more.

Figure 6.11   Tests for Several Related Samples window.

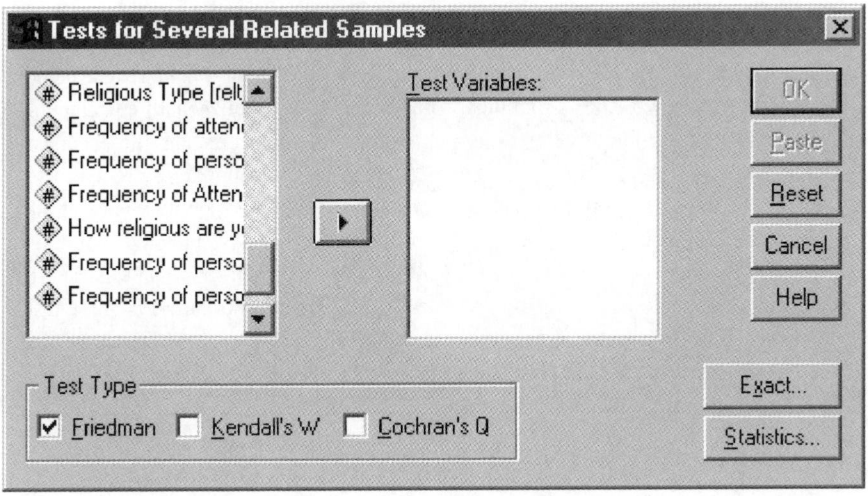

To illustrate the Friedman test in SPSS for Windows let us use the Frequency of personal prayer variable (how often the respondents report to pray). We have three measures of respondents' self-reported frequency of personal prayer: a 1st administration, and 2nd administration six months later, and a 3rd administration, one year after the first administration. Therefore we can use the Friedman test to examine whether individuals' frequency of personal prayer changes over the period of a year.

Pull down the **Analyse** menu and click on **Nonparametric tests...**, and then on **K related Samples....** You should now get a screen that looks like Figure 6.11.

Move the variable that represents the first administration of Frequency of personal prayer (freqpp) into the **Test variables** box. Then repeat this procedure for the second (freqpp2) and third (freqpp3) administration of Frequency of personal prayer. Make sure the **Friedman** box is ticked. Then press **OK.** You will then get the output shown in Table 6.13.

Within this output we have all the information needed to interpret whether a significant difference occurs between mean scores for administration of the measures of Frequency of personal prayer. Use *Describe* and then *Decide* to interpret this test. For this test, there are two stages we need to complete.

Using Describe and Decide to interpret the Friedman test (stage 1)

From the first two boxes in this output, you will need to consider three things:

- *Mean ranks* (Table 6.13, Ranks). These are the basis of our description. Here, we note the mean ranks. Higher mean ranks indicate higher scores on the variable.
- *The chi-square value* (Table 6.13, Test statistics). The statistical test statistic.

Table 6.13   Output for the Friedman test.

Ranks	
	Mean rank
Frequency of personal prayer	1.77
Frequency of personal prayer (6 months later)	1.91
Frequency of personal prayer (1 year later)	2.32

Test statistics[a]	
N	60
Chi-square	16.993
df	2
Asymp. sig.	0.000

[a] Friedman test.

- *The Asymp. Sig.* (Table 6.13, Test statistics). The significance level. This is the probability level given to the current findings. The significance level tells the researcher whether the difference noted between the means is different. Remember, if this figure is below the $p = 0.05$ or $p = 0.01$ criteria, the finding is significant. If this figure is above 0.05, the finding is not significant.

Therefore, the average mean rank for respondents on the first administration of Frequency of personal prayer is 1.77, the average mean rank for respondents on the second administration of Frequency of personal prayer is 1.91, and the average mean rank for respondents on the third administration of Frequency of personal prayer is 2.32. The significance level is $p = 0.000$. This is smaller than 0.01, therefore we conclude that there is a significant difference for mean ranks of Frequency of personal prayer over the three occasions.

Post-hoc comparisons with the Friedman test (stage 2)

Similar to the Kruskal–Wallis H test (and the ANOVA statistics), when a significant result is found for the statistical test it is necessary to break down the differences between pairs of means (1st and 2nd administration; 1st and 3rd administration; and 2nd and 3rd administration). Also similar to the Kruskal–Wallis H test, there is no option in the test screen for analysing these differences. Instead, researchers use the Wilcoxon sign-ranks test to compare these pairs of differences (remember the Wilcoxon sign-ranks test compares two sets of scores on a continuous variable that has been administered twice, that is suitable for use in a non-parametric test). So to break down the differences between administrations we need to perform a series of Wilcoxon sign-ranks test. Table 6.14 shows a series of Wilcoxon sign-ranks

Table 6.14 A series of post-hoc comparisons using the Wilcoxon sign-ranks test for Frequency of personal prayer by administration.

(a) 1st and 2nd administration

Ranks

		N	Mean rank	Sum of ranks
Frequency of personal	Negative ranks	9[a]	12.83	115.50
prayer (6 months later)	Positive ranks	15[b]	12.30	184.50
– Frequency of	Ties	36[c]		
personal prayer	Total	60		

[a] Frequency of personal prayer (6 months later) < Frequency of personal prayer.
[b] Frequency of personal prayer (6 months later) > Frequency of personal prayer.
[c] Frequency of personal prayer = Frequency of personal prayer (6 months later).

Test statistics[b]

	Frequency of personal prayer (6 months later) – Frequency of personal prayer
z	−1.089[a]
Asymp. sig. (2-tailed)	0.276

[a] Based on negative ranks.
[b] Wilcoxon sign-ranks test.

(b) 1st and 3rd administration

Ranks

		N	Mean rank	Sum of ranks
Frequency of personal	Negative ranks	7[a]	16.93	118.50
prayer (1 year later) –	Positive ranks	28[b]	18.27	511.50
Frequency of personal	Ties	25[c]		
prayer	Total	60		

[a] Frequency of personal prayer (1 year later) < Frequency of personal prayer.
[b] Frequency of personal prayer (1 year later) > Frequency of personal prayer.
[c] Frequency of personal prayer = Frequency of personal prayer (1 year later).

Test statistics[b]

	Frequency of personal prayer (1 year later) – Frequency of personal prayer
z	−3.354[a]
Asymp. sig. (2-tailed)	0.001

[a] Based on negative ranks.
[b] Wilcoxon sign-ranks test.

continued

Table 6.14   (*continued*)

(c) 2nd and 3rd administration				
Ranks				
		N	Mean rank	Sum of ranks
Frequency of personal	Negative ranks	8[a]	14.00	112.00
prayer (1 year later) –	Positive ranks	25[b]	17.96	449.00
Frequency of personal	Ties	27[c]		
prayer (6 months later)	Total	60		

[a] Frequency of personal prayer (1 year later) < Frequency of personal prayer (6 months later).
[b] Frequency of personal prayer (1 year later) > Frequency of personal prayer (6 months later).
[c] Frequency of personal prayer (6 months later) = Frequency of personal prayer (1 year later).

Test statistics[b]	
	Frequency of personal prayer (1 year later) – Frequency of personal prayer (6 months later)
z	−3.230[a]
Asymp. sig. (2-tailed)	0.001

[a] Based on negative ranks.
[b] Wilcoxon sign-ranks test.

tests comparing pairs of administrations for the Frequency of personal prayer variable.

A series of Wilcoxon sign-ranks tests are carried out to provide post-hoc comparisons of the mean ranks. For these tests we just use the figures in the **Test statistics** boxes, which indicate whether there are significant differences between the means. These tests show us that no significant difference occurs between scores on Frequency of personal prayer for the 1st and 2nd administration ($z = -1.089$, $p > 0.05$), but a significant difference occurs between scores on Frequency of personal prayer for the 1st and 3rd administration ($z = -3.354$, $p < 0.01$), and a significant difference occurs between scores on Frequency of personal prayer for the 2nd and 3rd administration ($z = -3.230$, $p < 0.01$). Examining the original mean ranks for the scales (1st administration = 1.77, 2nd administration = 1.91, and 3rd administration = 2.32), the present findings suggest that scores on the Frequency of personal prayer for the 3rd administration were significantly higher than scores on the Frequency of personal prayer for the 1st and 2nd administration.

Using Decide and Describe to report the Friedman test

The next stage is that you will need to report these statistics. There is a formal way of reporting the Friedman test. This comprises two elements. First, there is a formal statement of your Friedman test statistic; this must include:

- The test statistic. Each test has a symbol for its statistic. The Friedman test uses the chi-square statistic (χ^2). Therefore, in your write-up you must include what χ^2 equals. In the example above $\chi^2 = 16.993$.

- You must report the probability of the Friedman test. Again, this is done in relation to whether your probability value was below 0.05 or 0.01 (significant) or above 0.05 (not significant). Here, you use less than (<) or greater than (>) the criterion level. You report this criterion by stating whether $p < 0.05$ (significant), $p < 0.01$ (significant) or $p > 0.05$ (not significant). So in the example above, as $p = 0.000$, we would write $p < 0.01$ and place this after the reporting of the χ^2 value. Therefore with our findings, $\chi^2 = 16.993$, $p < 0.01$.

Second, if you found a significant difference for the Friedman test, then you report the pairwise comparisons. When reporting the series of Wilcoxon sign-ranks test statistics you just state the test statistic and whether p was less than (significant) or greater than (not significant) 0.05. This must then be incorporated into the text, to help the reader understand and conceptualise your findings.

In writing the text use the *Describe* and *Decide* rule to inform your reader of your findings:

- Remind the reader of the two variables you are examining.
- Describe which mean score is highest.
- Tell the reader whether the analysis of variance is significant or not.
- If the analysis of variance is significant, you must inform the reader where the differences lie between groups.

You can use all the information above to write a fairly simple paragraph, which conveys your findings succinctly, but effectively. Therefore, using the findings above you might report:

Frequency of personal prayer mean ranks were calculated for the 1st administration of Frequency of personal prayer (mean rank = 1.77), the 2nd administration of Frequency of personal prayer (mean rank = 1.91), and the 3rd administration of Frequency of personal prayer (mean rank = 2.32). A Friedman test indicated that a significant difference occurred between scores for Frequency of personal prayer for the three administrations ($\chi^2 = 16.993$, $p < 0.01$). A series of Wilcoxon sign-ranks tests were carried out to provide post-hoc comparisons of the mean ranks. These tests show us that no significant difference occurs between scores on Frequency of personal prayer for the 1st and 2nd administration *(z = –1.089, p > 0.05), but a significant difference occurs between scores on Frequency of personal prayer for the 1st and 3rd administration (z = –3.354, p < 0.01), and a significant difference occurs between scores on Frequency of personal prayer for the 2nd and 3rd administration (z = –3.230, p < 0.01). The present findings suggest that scores on the Frequency of personal prayer for the 3rd administration were significantly higher than scores on the Frequency of personal prayer for the 1st and 2nd administration.*

Writing a non-significant result for the Friedman test

Example: Frequency of personal prayer mean ranks were calculated for the 1st administration of Frequency of personal prayer (mean rank = 1.77), the 2nd administration of Frequency of personal prayer (mean rank = 1.91), and the 3rd administration of Frequency of personal prayer (mean rank = 1.02). A Friedman test indicated that no significant difference occurred between scores for Frequency of personal prayer for the three administrations ($\chi^2 = 3.993$, p < 0.05).

Again, you will come across different ways of writing tests, but you will find all the information included above in any write-up.

Calculating the Kruskal–Wallis H and Friedman tests by hand

In this chapter you have learnt a number of new skills. At this stage, asking you to learn how to do these tests by hand may be distracting from the main aims of the book. The reason for this is two-fold. First, is that working out these tests is a long, complicated statistic to calculate. Second, these are advanced statistics and due to the number of stages involved (e.g. using post-hoc comparisons statistics) it is prudent for you to start using SPSS for Windows.

References

Beit-Hallahmi, B. and Argyle, M. (1997). *The psychology of religious behaviour, belief and experience.* London: Routledge.

Fichter, J.H. (1952). The profile of Catholics religious life. *American Journal of Sociology,* **58**, 145–149.

Gallup, G.H. (1980). *Index to international public opinion.* Westport, CT: Greenwood Press.

Gerard, D. (1985). Religious attitudes and values. In M. Abrams, D. Gerard and N. Timms (Eds), *Values and social change in Britain.* London: Macmillan.

Gorsuch, R.L. and Venable, G.D. (1983) Development of an 'Age Universal' I–E scale. Journal for the Scientific Study of Religion, **22**, 181–187.

Yeaman, P.A. (1987). Prophetic voices: differences between men and women. *Review of Religious Research,* **28**, 367–376.

Answers to exercises

Exercise 1: Chapter energiser

The solution to the Minefield Puzzle is given in Figure 6.12.

Figure 6.12   Solution to Minefield Puzzle.

1	2	2	1				
2	x	x	1				
2	x	4	2	1			
1	1	2	x	1			
			2	2	2		
1	1	2	2	x	1		
2	x	2	x	3	2	2	1
x	2	2	1	2	x	2	x

Exercise 2: Assessment check for chi-square

The distribution of respondents by each of the cells shows that while females tend to be happy, males tend to be unhappy. A chi-square was used to determine whether there was a significant association between the two variables. A significant association was found between sex and self-categorisation of happiness ($\chi^2 = 2.603$, $p < 0.05$).

Exercise 3: Assessment check for Wilcoxon sign-ranks test

A Wilcoxon sign-ranks test was used to examine significant differences between self-rating of kindness before and after the film. A significant difference ($z = -4.544$, $p < 0.01$) was found for mean ranks for kindness levels before and after the film, with participants scoring significantly higher on self-rated kindness before the film (mean rank = 24.17) than on self-rated kindness after the film (mean rank = 15.00).

Exercise 4: Self-assessment check for the Kruskal–Wallis H test

Kindness scores were calculated, with mean ranks for the extrinsic group (mean rank = 29.58), the intrinsic group (mean rank = 24.54) and the non-religious group (mean rank = 23.08). No significant difference occurred between the groups' scores for how respondents described their kindness ($\chi^2 = 1.629$, $p > 0.05$).

CHAPTER 7

Summary and consolidating your learning

In this session there will be:

* a summary of the book;
* some exercises to consolidate your learning.

Exercise 1: Chapter energiser

Try the puzzle in Figure 7.1. This puzzle is designed to act as a reminder of all the concepts you have come across in the book.

Figure 7.1 Word Search Puzzle.

The 20 words relating to statistics have all been hidden in the grid. They have been printed across (backwards or forwards), or up and down, or diagonally, but always in a straight line without letters being skipped. You can use the letters in the grid more than once. You will probably find it helpful to mark the words in the grid.

N	O	M	I	N	A	L	B	E	I	H	T	X	I	W	C	V
P	O	R	N	A	I	J	L	L	M	N	I	I	R	O	O	S
T	I	N	D	E	P	E	N	D	A	N	T	S	N	U	R	T
V	T	P	P	M	E	A	P	D	D	U	U	T	P	Q	R	I
B	H	E	E	A	I	T	N	E	S	O	I	A	R	Y	E	U
S	C	G	S	P	R	E	R	R	T	N	R	L	O	C	L	G
I	C	N	I	T	P	A	I	G	U	A	R	A	B	N	A	N
G	B	L	H	E	U	M	M	O	M	C	C	A	A	E	T	O
N	R	A	D	Q	V	R	U	E	H	T	R	V	B	U	I	C
I	E	N	S	A	F	S	T	S	T	C	H	A	I	Q	O	E
F	O	I	M	K	Q	R	U	X	H	R	W	I	L	E	N	T
I	H	D	J	A	I	R	G	A	B	H	I	L	I	R	I	E
C	S	R	G	C	O	H	R	I	M	N	X	C	T	F	R	R
A	S	O	H	I	S	T	O	G	R	A	M	G	Y	R	O	C
N	O	I	T	A	I	V	E	D	D	R	A	D	N	A	T	S
C	X	A	U	P	K	A	A	V	I	A	N	O	C	Q	S	I
E	R	J	L	A	V	R	E	T	N	I	I	K	L	P	E	D

Summary

The main aim of the book was to view statistics as a decision-making process. We introduced you to this process in Chapter 1 through the decision-making chart (Figure 7.2).

In Chapter 2 we explored the nature of variables, in reference to Question 1 in the decision-making chart. To this end we identified three things:

- You can describe variables in two main ways: (a) categorical or (b) continuous.

- Each variable is made up of levels.

- Continuous variables can be described as demonstrating certain statistical properties that allow them to be used in *parametric* statistical tests. However, sometimes some continuous variables *do not* show these particular statistical properties, and when this happens, the variables are thought suitable to be used in *non-parametric* statistical tests.

We also showed you how to use SPSS for Windows to create a datafile.

In Chapter 3 you were introduced to ways of changing, combining and describing variables, both by hand, and in SPSS for Windows. We showed you ways of creating new variables from a number of items through the use of COMPUTE, and to manipulate variables using the RECODE statement. We also explored a number of descriptive statistics, such as frequencies, measures of central tendency (mean, mode and median), standard deviation, bar charts and histograms. Remember, we emphasised that descriptive statistics are used to *best* describe the data.

In Chapter 4 you were introduced to inferential statistics. In setting the scene for using the statistical tests described in Chapters 5 and 6, we outlined the importance of the concepts of the normal distribution, probability and significance testing. We also looked at the concept that is important to Question 2 in the decision-making chart. That is, whether continuous data can be viewed as suitable for use in either a parametric or non-parametric test.

In Chapters 5 (Parametric tests) and 6 (Non-parametric tests) you were shown how to perform a number of statistical tests, by hand, and in SPSS for Windows. Parametric tests included Pearson product–moment correlation coefficient, related *t*-test, independent groups *t*-test, analysis of variance – between subjects, and analysis of variance – within subjects (both with use of comparisons). Non-parametric statistical tests included the chi-square, Spearman's rho, Wilcoxon sign-ranks test, Mann–Whitney U test, Kruskal–Wallis H test and the Friedman test. In performing these tests there was an emphasis on remembering to *Describe* and *Decide* your findings.

Figure 7.2   Decision-making table for choosing statistical tests.

Question 1 What combination of variables have you?	Which test to use	Question 2 Should your continuous data be used with parametric tests or non-parametric tests?	Which test to use	Question 3 How many levels has your categorical data?	Which test to use
Two categorical	Chi-square				
Two separate continuous	Go to Question 2	Parametric	Pearson		
		Non-parametric	Spearman		
Two continuous which is the same measure administered twice	Go to Question 2	Parametric	Related *t*-test		
		Non-parametric	Wilcoxon sign-ranks		
Two continuous which is the same measure administered on three occasions or more	Go to Question 2	Parametric	ANOVA (within subjects)		
		Non-parametric	Friedman test		
One categorical and one continuous	Go to Question 2	Parametric	Go to Question 3	2	Independent-samples *t*-test
				3 or more	ANOVA (between subjects)
		Non-parametric	Go to Question 3	2	Mann–Whitney U
				3 or more	Kruskal–Wallis

Consolidating your learning

We will now seek to consolidate your learning by trying to bring a number of these elements together. We will do this by getting you to analyse some data. This dataset, named drivrisk.sav, is on both the disk that came with the book and available on the website (www.booksites.net/maltby).

This dataset is concerned with responses from 150 drivers regarding their driving risk-taking and hostile thoughts. The dataset also includes some information on the personal characteristics (sex, age, type of car they drive) of respondents. The data is laid out as follows.

Variable 1. SEX: Sex of the respondent. Within this dataset males are coded as 1, females are coded as 2.

Variable 2. AGE: Age of the respondent. Within this dataset age is coded as the number of years of the respondent.

Variable 3. CARTYPE: We have used the engine size of the car that respondents drive to create a variable. We have made three distinctions:

1. The respondent drives a smaller sized car (engine size anything up to 1500cc [1.5 litres]). Scored as 1.

2. The respondent drives a medium sized car (engine size from 1500cc [1.5 litres] to 2000cc [2.0 litres]). Scored as 2.

3. The respondent drives a large sized car (engine size over 2000cc [2.0 litres upwards]). Scored as 3.

Variable 4. DRISKTAK: Scores in the Driving Risk-taking Scale (Kidd and Huddleston, 1994). This is a self-report scale that contains 10 items representing aspects of risk-taking while operating a motor vehicle (Kidd, 2000). Example items include 'Take risks when driving' (Item 3) and 'Drive 5 to 10 mph over the speed limit' (Item 2). Respondents are asked to rate each item on a five-point scale: never (1), almost never (2), half of the time (3), almost always (4), and always (5). Possible scores range from 10 to 50. Higher scores on the scale indicate a higher degree of driving risk-taking.

Variables 5 (PHYSAGG), 6 (DERGOTH) and 7 (REVENGE) are derived from the Hostile Automatic Thoughts Scale (Snyder *et al.*, 1997). The scale is a 30-item, self-report scale that assesses the extent to which a person has automatic, hostile thoughts (Snyder and Yamhure, 2000). There are three subscales that tap different components of hostile thoughts:

- physical aggression (*Variable 5*: 11 items, e.g. 'I can think of a lot of terrible things I'd like to see happen to this person' [item 7]);

- derogation of others (*Variable 6*: 10 items, e.g. I hate stupid people [item 14]);

- revenge (*Variable 7*: 9 items, e.g. 'I want to get revenge' [item 25]). Respondents are asked to use a five-point Likert scale (Not at all = 1, to All the time = 5) to indicate the frequency with which they have experienced particular thoughts contained in the items during the past week.

Higher scores on each of the variables indicate a higher level of those sorts of hostile thoughts.

Using this information, carry out the following analysis that draws on all your knowledge and skills in the book. You will probably need to go back to parts of the book and revisit concepts/procedures outlined in each of the chapters. This is part of the skills we want to develop; use this book as a reminder to support you when you are doing statistical analysis in the future.

The analysis

Using the dataset:

1. Provide a description of your sample using the sex and age variables.

2. Use descriptive statistics (measures of central tendency and standard deviation) to show how scores on each of the four scales (driving risk-taking, physical aggression, derogation, and revenge) are distributed.

3. Use the skewness statistic to examine how scores on each of the four scales are distributed (driving risk-taking, physical aggression, derogation, and revenge). Think about any implications for choosing whether this continuous data is suitable for use in parametric or non-parametric tests.

Answer the following questions using SPSS for Windows. For each question you will need to use an inferential statistical test to answer the question.

4. Is there a significant difference between males and females in their driving risk-taking, physical aggression, derogation, and revenge?

5. Is there a significant association between sex of the respondents and the type of car they drive?

6. Is there a significant relationship between driving risk-taking and any of the other measures of hostile thoughts?

7. Is there a significant difference between the type of car the respondent drives and the levels of driving risk-taking?

In answer to each question, provide a full write-up of your findings.

Final comments

We hope you have found this book useful. You now have the basis for extending your knowledge and skills, through further reading, of the use of statistics and SPSS for Windows. Therefore, what you should be doing now is starting to explore the literature (through books and journal articles in your subject area) to develop an advanced understanding of statistics.

If you have completed this book, you will be ready to do this. You have experienced early success in statistics, and hopefully this will give you the confidence to continue. At times, statistics will seem complicated but do persevere, as you did with this book – you have taken on board a number of complicated ideas. Well done.

As a final note we encourage you to continue using the website (www.booksites.net/maltby). There are many things on these pages for you to continue to develop your knowledge and skills with statistics.

Good Luck in any further study you may do.

References

Kidd, P. (2000). The Driving Practices Questionnaire. In J. Maltby, C.A. Lewis and A.P. Hill (Eds). *A handbook of psychological tests*, Vol. 1 (pp. 188–190). Lampeter, Wales: Edwin Mellen Press.

Kidd, P. and Huddleston, S. (1994). Psychometric properties of the Driving Practices Questionnaire: assessment of risky driving. *Research in Nursing and Health*, **17**, 51–58.

Snyder, C.R. and Yamhure, L.C. (2000). The Hostile Automatic Thoughts (HAT) Questionnaire. In J. Maltby, C.A. Lewis and A.P. Hill (Eds). *A handbook of psychological tests*, Vol. 1 (pp. 200–205). Lampeter, Wales: Edwin Mellen Press.

Snyder, C.R., Crowson, J.J. Jr, Houston, B.K., Kurylo, M. and Poirier, J. (1997). Assessing hostile automatic thoughts: development and validation of the HAT Scale. *Cognitive Therapy and Research*, **4**, 477–492.

Answers to questions

Exercise 1: Chapter energiser

See Figure 7.3 for the solution to the Word Search Puzzle.

The analysis

1. There are 69 males and 81 females in the present sample. Age ranges from 18 to 46 years (mean age = 21.20, SD = 5.59).

2. Mean scores (with standard deviation in brackets) were as follows for each of the variables: driving risk-taking, mean = 29.53 (SD = 7.05); physical aggression, mean = 37.51 (SD = 19.69); derogation, mean = 28.38 (SD = 8.08); and revenge, mean = 27.37 (SD = 12.48).

3. Skewness statistics were as follows for each of the variables: driving risk-taking (0.183), physical aggression (0.186), derogation of others (−0.116), and revenge (0.516). As the skewness statistic is below the criterion of 1, we suggest this data is suitable for use in parametric tests.

Figure 7.3   Solution to Word Search Puzzle.

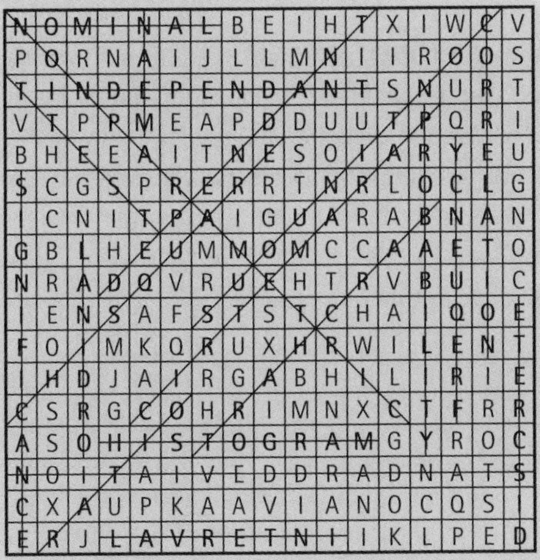

BARCHART	MEAN	
CHI SQUARE	NOMINAL	
CONTINUOUS	NON PARAMETRIC	
CORRELATION	ORDINAL	
DEPENDANT	PARAMETRIC	
DISCRETE	PROBABILITY	
FREQUENCY	RATIO	
HISTOGRAM	SIGNIFICANCE	
INDEPENDANT	STANDARD DEVIATION	
INTERVAL	T TEST	

4. Table 7.1 shows mean scores (standard deviations) for driving risk-taking, physical aggression, derogation, and revenge broken down by sex. An independent-samples t-test was used to examine differences between males and females for their mean scores on driving risk-taking, physical aggression, derogation, and revenge. No significant difference was found between males and females for their scores on physical aggression ($t=-0.914$, $p>0.05$), derogation ($t=-0.60$, $p>0.01$), and revenge ($t=-1.27$, $p>0.01$). However, males were

Table 7.1   Mean scores (standard deviations) of driving risk-taking, physical aggression, derogation, and revenge broken down by sex.

Variable	Males (N = 69)	Females (N = 81)	t value
Driving risk-taking	30.88 (06.3)	28.37 (07.5)	2.20*
Physical aggression	35.91 (19.4)	38.86 (20.0)	−0.91
Derogation	27.96 (08.3)	28.75 (08.0)	−0.60
Revenge	25.97 (11.8)	28.56 (13.0)	−1.27

* $p < 0.05$; ** $p < 0.01$

found to score significantly higher than women on the driving risk-taking measure ($t = 2.20$, $p < 0.05$),

5. Table 7.2 shows a breakdown of the distribution of respondents by each of the cells (with percentages in brackets). The frequency of respondents in each cell was evenly distributed with the greatest number of respondents falling in the female/middle car cell and the lowest number of respondents falling within the female/larger car cell. A chi-square was used to determine whether there was a significant association between the two variables. No significant association was found between sex and car type ($\chi^2 = 1.973$, $p > 0.05$).

Table 7.2 Breakdown of 150 respondents by sex and type of car they drive.

		Car type		
		Smaller (<1500cc)	Middle (1500–2000cc)	Larger (>2000cc)
Sex	Male	29	23	17
	Female	32	35	14

6. A Pearson product–moment correlation coefficient was used to examine the relationship between scores on the Driving Risk-Taking Scale and scores on each of the measures of hostile thoughts. A significant positive correlation was found between driving risk-taking and physical aggression ($r = 0.208$, $p < 0.05$), derogation ($r = 0.266$, $p < 0.01$), and revenge ($r = 0.231$, $p < 0.01$), suggesting that there is an association between taking risks when driving and all three aspects of hostile thoughts.

7. Driving risk-taking scores were calculated for the smaller sized car group (mean = 27.43, SD = 5.4), the medium sized group (mean = 28.12, SD = 5.9), and the larger sized car group (mean = 36.29, SD = 7.8). An analysis of variance indicated a significant difference between the groups on the measure of driving risk-taking (F(2, 147) = 23.627, $p < 0.01$). A series of Scheffe tests revealed that respondents with larger sized cars scored significantly higher on driving risk-taking than both respondents with medium sized cars ($p < 0.01$) and respondents with smaller sized cars ($p < 0.01$). However, there was no significant difference between respondents with medium sized cars and respondents with smaller sized cars for scores on the measure of driving risk-taking ($p > 0.05$).

Index

Note: **page numbers** in bold indicate major topics and tests. **Words** in bold indicate commands, menus and boxes in SPSS for Windows